JOHN VINCE

D0519514

Discovering
Windmills

SHIRE PUBLICATIONS LTD

Contents

The cover photograph shows Skidby tower mill, Humberside.

ACKNOWLEDGEMENTS

Photographs are acknowledged as follows: Mark Arman, plate 4; Friends of Shipley Windmill, plate 13; Ray Harlow, plate 3; Peter Hill, plate 1; Cadbury Lamb, plates 2, 5, 6, 7, 8, 12, 14, 15, 16, 17 and front cover; Museum of East Anglian Life, plate 11; Norfolk Windmills Trust, plate 9; John Vince, plates 18, 19; Wrates Studio, Skegness, plate 10. The line drawings are by the author; the drawings on page 19 are from his book *Power Before Steam,* published by John Murray.

The author and publishers are grateful to the many mill owners and custodians who have provided information and assistance in the preparation of this book. Vincent C. Pargeter provided the dates for the Kent mills in the gazetteer and Cedric Greenwood provided the lists for Cheshire, Lancashire and Merseyside.

Set in 9 point Times roman and printed in Great Britain by C. I. Thomas & Sons (Haverfordwest) Ltd, Press Buildings, Merlins Bridge, Haverfordwest, Dyfed.

1. Introduction

Few things add as much atmosphere to the English country-side as a windmill. Even in decay a mill possesses a dignity few other buildings can equal. The history of windmills in England stretches back to the turbulent days of the twelfth century. One tradition suggests that they were introduced into Britain by Crusaders returning from the wars. This may be partly true, but all that we can say with certainty is that windmills were first built in Britain some eight centuries ago. None of the original structures remains; but among the mills which have survived to the present are some that are at least 300 years old and many more over a century.

In feudal times villagers were compelled to take their hard won grain to the Lord of the Manor's mill to be ground. The miller took a toll for his work, and it is easy to see how the traditional mistrust of the unfortunate miller came about. As Chaucer observed—

> His was a master hand at stealing grain.
> He felt it with his thumb and thus he knew
> Its quality and took three times his due . . .

The number of windmills still at work today is a minute fraction of the many hundreds which once helped to grind the nation's grain. When steam power was applied to flour milling in the middle of the last century the decline of the windmill began. Steam power could be generated at will and had many advantages over the wayward wind. As a result, millers introduced steam plants to supplement their mills' grinding capacity, and when a mill itself reached the point where it required expensive repairs the steam mill supplanted it.

The final blow to the windmillers' craft came when steel-roller mills began to dominate the flour trade in the 1880s. From then onwards the miller was sometimes reduced to grist milling alone, and it is not surprising to find that many mills were finally forced out of business just after the 1914-18 war.

During the 1930s redundant mills of all types began to decay. Fortunately at this time the Society for the Preservation of Ancient Buildings (SPAB) began its campaign to protect windmills. Many were still working and many were saved.

Windmills represent an important part of our technological history; and while it may not be possible or desirable to try to preserve every mill there are many sound reasons in favour of ensuring the survival of some. Posterity is entitled to a share of the heritage we may have either ignored or enjoyed.

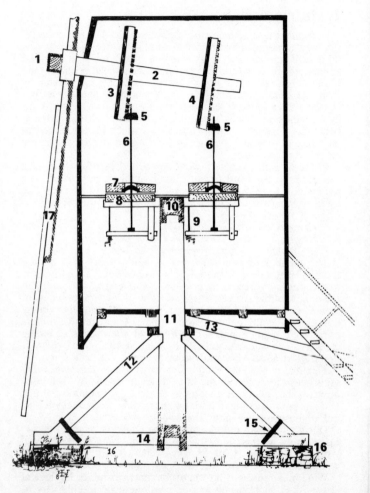

Overdriven post mill diagram. Gear train: *1. Poll End. 2. Windshaft. 3. Brake wheel. 4. Tail wheel. 5. Wallower. 6. Quant or crutch pole. 7. Runnerstone. 8. Bedstone. 9. Tentering gear.* **Structure:** *10. Crown tree. 11. Main post. 12. Quarter bars. 13. Tailpole. 14. Cross tree. 15. Strap to inhibit movement of quarter bar. 16. Masonry pier. 17. Sailstock and whip.*

2. Post mills

Although the earliest mills have long since disappeared we can say with some degree of assurance what they looked like. Representations of windmills have survived in manuscript records, carvings, stained glass, plasterwork and paintings.

One of the oldest windmill pictures is an engraving (*c* 1349) on a memorial brass in the church of St Margaret's, King's Lynn, Norfolk. Good carvings in wood (*c* fifteenth century) can be seen on a misericord at Bristol Cathedral, and on sixteenth-century bench ends in the parish churches of Bishop's Lydeard and North Cadbury, Somerset. A fine fifteenth-century windmill picture also survives among the stained glass of Fairford church, Glos. Each of these illustrations shows the simplest type of windmill — the post mill.

The wooden framed body of the post mill was supported by a massive upright post. This was held in position by quarter bars which rested upon horizontal cross trees. Most post mills had two cross trees at right angles to each other; and the ends of these were usually supported by short pillars of masonry which allowed the air to circulate freely and keep the timbers dry. As the diagram shows, the post did not rest upon the cross trees but was suspended a little above them.

There were, however, at least eight post mills which had three cross trees instead of the normal two, and six quarter bars in place of the usual four. The last mill of this kind stood, until late 1967, at Chinnor, Oxon. Its tottering frame was dismantled to make way for a house. The other mills which had the same type of trestle were at Wheatley, Oxon; Bledlow Ridge (demolished *c* 1933) and Stokenchurch (collapsed 1926), both in Bucks; Moreton, Essex (demolished 1964); Scotter, Lincs; Costock, Notts and Syston, Leics.

A mill which retains the simple lines of the early post mills can be seen at Bourn, Cambs. It has plain gabled ends and an open trestle. The precise age of this mill is not known but it must rank among the oldest surviving specimens.

The mill which can, at present, claim to be the oldest dated mill in England is at Pitstone, Bucks. One of its timbers bears the date 1627. The mill was given to the National Trust in 1937 by its former owner Mr L. J. Hawkins. Its working life ceased when it was badly damaged in a severe storm in 1902. This splendid mill attracted the attention of a group of local enthusiasts who embarked upon an ambitious scheme to restore the mill to its former grandeur. After a sustained effort this must rank among the best kept mills in England. Yet not many years ago it had been in danger of extinction.

In common with most timber framed buildings the members of a post mill usually have the principal joints numbered. This

5

made erection an easier operation especially if the timbers were prepared in one place and transported to another for assembly. Roman numerals were often used to mark joints, but other symbols, based on the Runic characters, were also employed by some craftsmen.

Although some post mills still have an open trestle most have had a brick roundhouse added to provide extra protection against the elements. In turret post mills the body of the mill turns on rollers on the top of the roundhouse as well as on the post.

One of the miller's daily tasks was to adjust the mill's position so that the sails faced the wind. To make this task easier post mills have a long tiller beam or tailpole which projects from below the superstructure at the rear. An added refinement was a cart wheel at the end of the beam. Whem mills of this type were in use a series of posts was arranged around the circumference of the circle described by the tiller beam to provide fixed anchor points. Another method of revolving the carefully balanced mill was to harness a horse to the tailpole.

With the progress of time a mechanical and automatic method was introduced to enable the mill to adjust itself to any change in wind direction. The device was a fantail and for post mills there were three different methods of attaching them. A few mills—as at Nether Dean, Beds—had fantail gear attached to the mill body above the rear gable (this mill was really a hybrid). Such a method does not seem to have been the most popular, probably because the driving mechanism was not readily reached for repair and maintenance. The second method employed a separate framework or carriage—as at Cross-in-Hand, Sussex—which was fixed to the tailpole. A fan carriage usually had two iron wheels at ground level. The third way was to erect a suitable framework upon or above the mill ladder. The two latter methods were the most popular, and they both required a rail track or paved circular path around the mill to assist their passage. It is interesting to note that the automatic fantail was not extensively adopted in Europe.

Another device sometimes fitted to the tiller bar was a simple lever—a talthur. A short chain was attached to this and to the lower part of the ladder. When the mill had to be luffed, to move the sails into the wind, this lever enabled the miller to raise the steps clear of the ground.

Many post mills were extended during their working careers, but they were designed in such a way that the only possible form of extension had to be on the rear gable forming a simple lean-to. Among those with this sort of extension were Argos Hill, Sussex, and Great Chishill, Cambs.

Post mills originally had common sails (see the later chapter on sails), but in time a large number of them were modernised by the substitution of patent or spring set sweeps.

Their original gearing was entirely wooden, and when it later required replacement cast iron cogs were frequently used.

In many ways the post mill is perhaps the most picturesque of our English windmills. Even when they begin to totter their grace and dignity is not lost. Forgotten and isolated, a number of mills struggle for their existence, but year by year they slowly bend to the will of the wind. Too soon they will be gone, and men will forget where they once stood.

Other mills have been more fortunate, and for those which have attracted the attention of private or public benefactors we should be grateful. A mill restoration usually depends upon local dedicated enthusiasts, and without the efforts of many such groups the English countryside would be very much the poorer. Few things can complete a prospect so well as a windmill—and where proportion and elegance are concerned other mills, for all their charm, cannot equal the lines of a post mill.

One of the taller post mills undergoing restoration is at Framsden, Suffolk. This fine mill, with its ogee gable, dominates the village skyline. Few can fail to admire it as it soars above the cottage pantiles. East Anglia is a Mecca for the windmill hunter, and one of its treasures is undoubtedly the mill at Saxtead Green, which is open to the public. The mills at Friston and Thorpeness are also excellent examples and follow the same Suffolk tradition.

The splendour of these working mills impressed William Cobbett as long ago as 1830. Writing about his approach to Ipswich, he noted that the windmills on the hills . . . are so numerous that I counted, whilst standing in one place, no less than seventeen. They are all painted or washed white; the sails are black; it was a fine morning, the wind was brisk, and their twirling altogether added greatly to the beauty of the scene . . . and . . . appeared to me the most beautiful sight of the kind that I ever beheld' *(Rural Rides)*.

Not all mills were painted in the manner Cobbett describes; in Buckinghamshire, for example, there seem to have been only two like this. One of these stood at Great Missenden and was burnt down in 1876. Fortunately the artist E. J. Niemann recorded it in a fine landscape he painted in 1868.

3. Smock mills

The alternative to turning the entire mill body to face the wind was to provide the mill with a revolving cap to which the sails were fixed. History has neglected to record the name of the genius who invented the method of attaching the sails and windshaft to a mobile unit that was both lighter and easier to manipulate, but his innovation allowed mill design to evolve a stage further.

The development of the smock mill represents the second stage in windmill design. Smock mills, so we are told, derive their name from an alleged resemblance to a man in a smock. Whatever the truth of the matter the name has stuck, with the variation 'frock' mill, and wooden framed tower mills retain it to this day.

Post mills are constructed on a rectangular unit, but the smock mill design is derived from a circle. Wooden members, which had to be straight, reduced the circle to a series of connected chords. The usual pattern of smock mill construction was based upon eight sides, but six, ten and twelve sided mills have also been constructed. This departure from square jointed construction must have produced problems for the millwrights to solve. The key structural members of a smock frame are the stout cant posts which form each corner. It must be remembered that these posts are closer together at the top than they are at the base. To form good joints that were leaning in two dimensions must have called for all the carpenters' skill. To prevent the base of the cant posts from rotting too quickly the smock towers were often raised on a brick plinth. This had the effect of prolonging the life of the posts, but in time many mills ran into serious difficulty when a corner post became weak and threatened the stability of the whole structure.

The diagram opposite shows the main structural members of a typical smock mill. One of the essential needs of a smock and a tower mill is a level and stable curb upon which the cap may rotate freely. From the way the cant posts are arranged they naturally tend to want to move outwards and allow the structure to collapse. Even small movements within the frame could allow the curb to buckle out of true, and anything which hindered the free movement of the cap was an anxiety to the miller. Perhaps the greatest hazard arose when a cap jammed and the wind direction changed suddenly. If the wind was very strong and blew from behind the sails, making the mill tail-winded, the consequence could be very serious indeed. A tail wind could blow down a post mill, or remove a smock mill's cap and sweeps or sails in far less time than they would take to replace.

In some ways smock mills are far more vulnerable than post mills. They have sloping sides and usually twice the number of corners, if not more; these all provide additional chances for the elements to penetrate and destroy. Joints in the weatherboarded cladding were sometimes sealed with zinc or lead strips, but in spite of these precautions the miller always needed to keep a weather eye on his mill. The provision of windows in the sloping sides of a smock mill also gave rise to problems, and

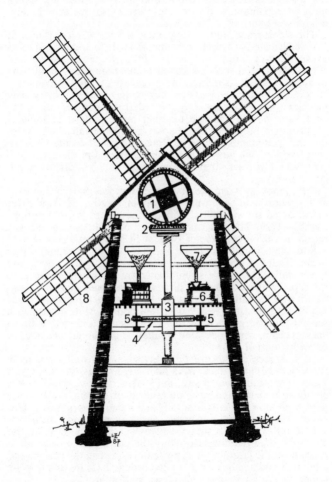

Diagram of tower mill with underdriven stones: *1. Brake wheel. 2. Wallower. 3. Vertical shaft. 4. Spur wheel. 5. Stone nuts. 6. Stones within vat. 7. Hoppers for grain. 8. Common sails.*

where the window frame joined the body special care was needed to make it weatherproof. Like the post mills some smock mills were painted, and were therefore expensive to maintain; others were tarred.

The oldest smock mill in England is at Lacey Green, Bucks. It was constructed at Chesham round about 1650 and moved to its present site in about 1814. Most mills suffer from a move, but the Lacey Green mill has lasted better than many built long after. Its recent condition was very poor and a complete restoration is now being carried out by the Chiltern Society. The machinery is almost certainly the oldest wooden gearing in the land, and its size and age together must make it of unusual archaeological interest. Quite apart from its technical attributes this mill commands one of the finest prospects in the Chiltern countryside; and now it is protected from the elements and refurnished with sails it adds more than a touch of dignity to the landscape.

We cannot say when the first smock mill was built in England, and it seems as if they were actually invented in Holland. The oldest dated smock mill is probably the one at West Wratting, Cambs—1726. The last smock mill to be built in England was raised in 1928-9 by Homan Bros at St Margaret's Bay, Kent.

Some smock mills employed a counterpart of a post mill's tiller beam. Tailpoles were a common feature on many smock and tower mills. They were strengthened by outriggers attached to the cap. To assist the luffing process some smock mills were also fitted with fantails. These were lighter structures than those required to turn the heavy post mill bodies. Mills often had a second method of rotating the cap—usually effected with an endless chain, which operated a pulley on the fantail mechanism, or with internal gears.

Early smock mills were probably, like post mills, originally fitted with common sails. At a later stage spring or patent sails were substituted on many mills, but a few smock mills demonstrated the innate conservatism of certain millers who used both a pair of common sails and a pair of spring or patent sweeps—as at Earnley, Sussex, and Lacey Green, Bucks.

If a mill was made with a tall tower—perhaps to avoid the effects of other buildings close by—the sails might be too high to reach from the ground. So a new feature arose in the design, and a tall mill had a gallery constructed around its sides to allow the miller to attend to his canvas, as at Cranbrook.

Smock mills often had a tarred base, and they always seem particularly attractive if their superstructure is painted white. A number of smock mills have survived to adorn the countryside, and some of them serve as homes.

A striking example is Ibstone, Bucks. In 1967 a mock fantail and sails were temporarily added to it for effect by a film company which used the mill for some location shots in the film *Chitty Chitty Bang Bang*. Whatever the purists may say about mock sails there can be no doubt that the mill looked all the better for them. It finished work in 1914-18, and its present condition suggests that it will last for many years to grace the skyline above the picture-book village of Turville. The mill was formerly owned by St John's College, Oxford.

4. Tower mills

The difference between a smock mill and a tower mill depends upon the materials used. A smock mill has a wooden frame, and a tower mill is usually constructed from stone or brick. Both have the same basic shape and are arranged internally on the same principles. The slope of the mill wall is called the batter. Most towers are round in plan, and it is unusual to find octagonal towers as at Wheatley, Oxford. Another variation was to make the lower part of the tower octagonal in plan and the upper portion round, as at Great Bardfield, Essex.

Tower mills represent the pinnacle of mill design. They were probably introduced at about the time of the first smock mills and, generally speaking, are of larger proportions than other types. Various materials have been used to protect them, and many stone towers were enhanced by a veneer of tiles or slates. The wall of the tower mill was usually very thick and, like some of the smock mills mentioned above, a good number of them have been converted into homes. The octagonal brick tower at Wendover, Bucks, has long been inhabited and its present owner still lives in comfort under the largest mill cap in England. Stone or brick towers were easier to weatherproof than the boarded smock mills. The owner could employ a cement render if the brickwork began to crumble. This was not always visually pleasing, but it answered the miller's purpose. One other safeguard was to paint the tower with tar. A good coat could last for many years and it was a sound practical answer to the driving rains of an English winter. Some of the most attractive towers are those with tarred sides and white paintwork, as at Heckington, Lincolnshire.

Tower mills had many advantages over post and smock mills. One poor feature, however, arose in towers which had windows and door openings placed one above the other. This arrangement allowed lines of weakness to develop if the foundations subsided even a little. A superior and safer method was to place openings around the body in a spiral fashion. Almost any tower mill with vertically spaced openings shows signs of some subsidence in its structure. A dropped window arch could easily lead to the distortion of the curb and this all too often made it impossible to revolve the cap, an expensive thing to correct. In some circumstances a jammed cap could mean the final closure of the mill.

Like smock mills the tower mills were provided with common, spring or patent sails according to their age and prosperity. Most of them had a gallery and even those that used common sails until the end of their days usually had a fantail mechanism. The finest tower mills were built in the late eighteenth century and early nineteenth. This was the great age of tower building, but even late in the nineteenth century some millers were wealthy enough to restore, refit or rebuild their mills.

In many ways tower mills are more durable than their wooden counterparts, but they can disappear without trace—like the flint rubble tower at Holmer Green, Bucks, which collapsed in 1929 after a century of life. This mill was unusual in this part of the country—it had a triangular gabled cap which was turned by an endless chain on a pulley, after the style of Melin-y-Bont, Gwynedd, and mills on the Fylde, Lancs. Mills built of coursed stone or brick stand more chance of survival when they fall into disuse. A great number of long redundant towers still almost intact add interest to the English countryside. Others for one reason or another survive in a truncated form. The remains of Great Horwood mill, Bucks, stand, inhabited by pigs, behind a modern barn. The superstructure was dismantled at the beginning of the last war when an airfield was built close by and the tower was considered to be a hazard to aircraft. A few miles away the stump of Stewkley mill also survives; it was taken down to its present level in 1922 when its upper part was in a dangerous condition.

A study of Ordnance Survey maps suggests that when a mill is partly dismantled it is then, quite properly, omitted from the next revision of any given map sheet. The Second Series of the OS 1 : 50,000 maps does not distinguish between mills in use and those that are disused, nor are partial remains shown, and the diligent searcher will need to use an older edition as well as a new one if he is to get the best results from his wanderings. Post and smock mills do not often leave many visible traces behind when they finally fall to the ground, but the base of a tower mill may quite often survive behind a hedgerow or the miller's house.

All three types of mill have been erected on man-made mounds which were constructed to add effective height to a mill structure. On crowded sites where, perhaps, large barns or other buildings could affect the free passage of the wind such an arrangement was a satisfactory and less expensive alternative to the erection of a high structure. Where a mill has completely vanished a mound will often remain as it is a costly item to eradicate. In remote places known mill mounds have been slowly, over many years, destroyed by the plough. A good specimen of a mill mound with its tower may still be seen at Blackthorn, Oxon.

5. Drainage mills

The work of draining the Fens began as far back as the sixteenth century. Wind power was set to work to aid man's efforts to reclaim valuable areas of land and from this need the drainage mill evolved.

Earlier drainage mills were plain smock structures, but later specimens were built of brick and were often tarred. Drainage mills usually had a Dutch look about them with their dumpy towers, common sails and tailpoles. In fact, the design seems to have originated in Holland. Latterly patent sails and fans replaced the tailpole.

Internally their machinery was simple. The wind turned the sails and the windshaft. A wallower, fixed to the vertical shaft, was in turn driven by the brake wheel, and so the shaft was rotated. At ground level a bevel gear was connected to the pit wheel, which was mounted on the same axle as the paddle wheel. The revolving paddle wheel scooped up water which it discharged at a higher level.

Marshmen rarely lived with their families in remote mills. Many were so isolated that the only way to reach them was by water. The decline of the drainage mill is closely linked with the decline of its corn grinding counterpart. Steam pumps came into use early in the nineteenth century. Many drainage mills continued to work until the 1930s. They gradually decreased in number and although one was built as late as 1912 we can no longer number them in hundreds.

Some fine looking mills remain, however, among them Stracey Arms, near Acle Bridge; and High Mill, Berney Arms, Norfolk (preserved). The latter mill is a very interesting structure. Its scoop wheel is detached from the mill, and a long shaft provides a link between them. This was one of the few drainage mills which performed a milling operation—grinding cement clinker. Other drainage mills in various places were fitted with stones for corn grinding, but mills of this type were the exception rather than the rule.

6. Hybrids

Like species in the natural world, windmills have also produced their hybrids. There were a few mills which were half post and half tower. The reason for their existence may be explained by a reluctance to demolish the sound body of a post mill when its working life could be extended by placing it upon a short tower. In such cases the gabled framework was mounted on a squat tower which had a curb like a conventional tower mill. This allowed the whole body to be luffed into the wind as conditions required. A fantail—above the rear gable—and a tiller beam of the usual post mill pattern provided alternative methods of moving the sails into the wind. Every rule has its exceptions, of course, and the composite mill at Monk Soham, Suffolk, was in fact a purpose built structure and not the result of remodelling. Composite mills are also recorded at: Banham and Thornham, Norfolk; Rishangles, Suffolk, and Cowick, Yorks.

Another variation on the mill design was to place a reduced post mill body on to a smock tower. The driving power was transferred from the windshaft by a shaft passing through the vertical post, which was connected to the other machinery below. Not many of these hollow-post mills seem to have been built in England, and the one on Wimbledon Common is a rebuilt survivor. The lower part of this unusual mill has been converted into a residence. Its sails were in proportion to the post mill style cap. This adjusted itself to changes in wind direction by the fantail—situated on an outrigged frame behind the rear gable. The original building is said to date from 1817, and its present appearance results from a rebuilding in 1890. Apart from its technical distinctions the mill house has an interesting literary association. Lord Baden-Powell wrote part of *Scouting for Boys* in the adjacent miller's cottage in 1907. Two original hollow-post mills remain at Stodmarsh, Kent, and Acle, Norfolk.

7. Caps

The Mill Cap

The crowning glory of the tower mill was its cap and sails. Over the years a good many disused mills have lost their caps. Some were purposely removed, others decayed and collapsed, or were forcibly removed by winter gales. Next to the sails the cap and its fittings are the most vulnerable feature of a mill. Cap design can tell us so much about a mill and it is a vital detail to record.

Some writers have suggested that certain styles of caps are common to given geographical areas. To some extent this is true but any general pattern of distribution tended to be affected by migrant millwrights, who sometimes went a long way from their homes to undertake work, and as a result produced a mill with features that were different from those of its immediate fellows. An example of this sort occurred at Much Hadham, Herts, where a millwright from Louth, Lincs, constructed an elegant eight sailer.

The function of the cap was to protect the tower and the machinery immediately below its roof. One of the most difficult areas to secure against the elements was the weather beam and forward end of the windshaft. Another important sector was the curb on which the cap revolved. This was usually guarded by a deep board or in the case of conical caps a neat and deep petticoat tastefully scalloped.

We can distinguish six main types of cap which may be found on smock and tower mills alike. They are: the gabled, the post mill type, the cone shaped, the domed, the ogee and the upturned boat shape. It would be difficult to assign any definite periods of time to these different styles. They probably represent no more than a variety of solutions to the same technical problems.

The Gable Shape

One of the simplest forms of cap, and probably the earliest, is the triangular gable and its variations. Examples of this kind could be found in widely scattered parts of England. They appeared as far north as the Fylde, Lancs; Wirral, Merseyside; Lutton Gowts, Lincs, and Gwynedd. In the south they could be found at Bembridge, Isle of Wight; Holmer Green, Bucks; and at Ashton, Somerset. It is interesting to note that what is England's only thatched mill—at High Ham, Somerset—also has a cap with contours that follow the swept gable pattern.

The Post Mill Shape

Many mills in Kent, Essex, Surrey and Sussex had caps shaped like miniature post mills. A sketch dated 1796 shows a Lincolnshire mill with a cap of this type, and the design may have been more general than present evidence suggests. Notable examples of this are at Headcorn, Canterbury and Cranbrook, Kent; Rye, Chailey, Staplecross and West Chiltington, Sussex. Caps of this kind do vary considerably in their proportions. Some are very deep and appear to sit heavily on their towers, others have a sleeker look and seem almost to float on air.

The Boat Shape

Another, East Anglian, type of cap has a remarkable similarity to an upturned dinghy. This style is related to the post mill roofs but with one important difference. In plan its base timbers, instead of being parallel, are curved to match the contours of the upper part of the mill tower. Caps of this kind are in two forms: those with gables of almost equal proportions and those with a larger front gable and a more pronounced sweep to their ridge profiles.

Caps with balanced gables can still retain their individuality. If the cap is boarded in the usual way it can have quite a sleek look. An alternative can be found among those drainage mills where the weatherboarding is bound with additional strips running at right angles. This gives the impression of half a beer barrel, and has a much heavier appearance. Smock and tower mills may be found with boat shaped caps; and the most pleasing version visually is probably the one with the dramatic line to its ridge—like Gibraltar Mill, Great Bardfield, Essex.

Conical Caps

The most elementary alternative to a conventional rectangular roof protecting a round tower is to provide a cone shaped cap. This idea seems simple enough but it demanded a greater degree of skill to construct. Caps of this kind do not seem to have been as common as other types—perhaps because they were re-designed when they required extensive repairs. Conical caps are either based on a circular plan or on a hexagonal or octagonal plan, as on the smock tower of King's Mill at Shipley in Sussex. Another mill with an almost identical cap may be found at Cley-next-the-Sea, Norfolk.

Domed Caps

A dome is a logical development from a cone or pyramid, and its construction demands a considerable degree of skill. In Sussex they were shaped rather after the fashion of beehives and their proportions were light in appearance. Examples of typical Sussex domes are Polegate and Selsey. One at Patcham had the roundest of all.

In East Anglia the caps mostly sit heavily on squat towers—like deep bonnets on rotund matrons. The Cambridgeshire mill at West Wratting is typically heavy in appearance. Although the Suffolk smock mill at Alderton had a taller tower than most, it too boasted a large bonnet. It is the cap which gives the mills in this part of England their distinctive look and makes them readily distinguishable from their counterparts in the south and even those further west in Bedfordshire.

Going westward, Wilton Mill, Wilts, sports a dome. It comes in the East Anglian class, but the profile is rounder and gives an elephantine impression.

The Ogee Cap

The ogee curved dome is the final glory of tower mill design. Among the ogee caps of England, those of Lincolnshire justly claim pride of place. Here they abound *par excellence* and in one form or another their influence extends across the eastern and midland parts of England. Sussex, too, has a few caps that are rather blunt specimens of the ogee style. The most noted one, Halnaker, was recorded by William Turner and his painting may still be seen in the National Gallery of Scotland.

Another example is the mill at Ewhurst, Surrey. Lurid tales are woven around this mill's working days when it appears to have been prominently involved in the distribution of contraband on its way from the coast to London. After the mill ceased work it was converted into a residence.

North of the Thames is the real ogee land, and many fine examples lie between the Thames Valley and Yorkshire. Out in the open Oxfordshire countryside—at Milton Common and Wheatley—we may see examples of the low pitched ogee cap which is picturesque without being elegant. Some caps in this part of the world have a copper cladding which weathers a distinctive green. Across the Oxfordshire border limestone mills are not unusual and in addition to the two mentioned above we may add Blackthorn—which has a tile hung tower.

Two distinctions remain to be made about ogee caps. Those in the Cambridgeshire region—as at Burwell—have a slightly flared emphasis at the base, and in Lincolshire the fashion is generally in favour of an inward curve which provided an unmistakable onion look. The details of finial decoration vary but a round knob seems to be the favourite. South Midland finials tend to be heavier in design, and in Oxfordshire an acorn style is used. Yorkshire caps, which may well have been the first in ogee style, are also quite distinctive.

Materials

Millwrights have employed many different materials in their efforts to weatherproof their mills. We may guess that the first caps on the early towers were thatched, but they could equally well have been boarded. Although corrugated iron adorns many a cap these days, millwrights have made good use of a variety of materials in the past. Chesterton, Warwicks, is leaded; copper was popular in Oxfordshire but perhaps the most universal alternative to these was tarred canvas on boards (called marouflage finish), which has many enduring qualities and can be very attractive if it is skilfully applied—as at Heckington, Lincs.

8. Sails

Common Sails

Early windmills had simple rectangular wooden frames upon which the miller arranged his canvas. Sails were constructed around two sailstocks set at right angles and mortised through the windshaft. Mills constructed in later times, about the seventeenth century, had the lengths of the sails increased by the addition of whips (lighter timbers bolted to the sail arms). An important difference between the primitive sails and those in use in the early eighteenth century is shown in the diagram (page 19). The canvas on the latter was rigged on a single sided blade in place of the earlier two sided arrangement. One of the important developments in windmill design came about in the 1750s when John Smeaton introduced the use of cast iron to supplement the traditional timber. When this happened it became possible to design cross arms and canisters to take stouter sailstocks; and iron gears were introduced to replace or complement existing wooden cogs. The use of iron gears running next to wooden wheels was said to make for quieter running and less vibration. The great disadvantage of common sails was the necessity to stop the mill working when the wind-strength changed so that the area of canvas could be adjusted. This was a tiresome and, in winter or stormy weather, a dangerous task.

Spring Sails

In 1772 a Scottish millwright—Andrew Meikle—invented a new type of sail which was composed of a series of shutters arranged like a Venetian blind. The shutter blades were opened or closed depending upon the state of the wind and how much work the miller had to do. In order to adjust the sails it was still necessary to stop the mill so that the bars connecting the rows of shutters on each sweep could be moved to the required position. Each sail was still altered independently, but the adjustment was a much easier matter than dealing with freezing canvas while clinging to a sweep high above the ground. The angle of each row of shutters was controlled by the pressure, transmitted by the shutter bar, from a powerful spring. Sails of this type were known as spring sails, and many mills had double rows of shutters on their sweeps.

Roller Reefing Sails

A variation of Andrew Meikle's system was patented in 1789 by Captain Stephen Hooper. He replaced the shutters with a roller blind system which was claimed to be automatic, but in practice was remotely controlled by the miller. This type of sail, with its distinctive 'V'-shaped air poles to control the shutters, was once a feature of many mills in Yorkshire in particular. The

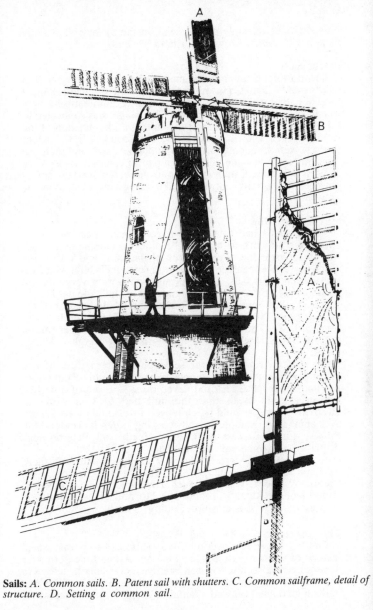

Sails: *A. Common sails. B. Patent sail with shutters. C. Common sailframe, detail of structure. D. Setting a common sail.*

only surviving example can be seen on the tower mill at Bally Copeland, County Down, Northern Ireland.

Patent Sails

The most significant improvement in sail design came about in 1807 when William Cubitt invented his patent sail. This retained Meikle's shutters, but they were controlled automatically by a weight suspended outside the mill. This weight was connected to the shutter bar by a series of rods linked, at the junction of the sails, to a lever known as the spider. Like spring sails, the patent shutters were frequently arranged in double rows; and their use was combined on some mills with a pair of common sweeps—as at West Wratting, Cambs. A later advance combined the spring device with a patent control to make spring sails fully automatic as well.

Multi-Sailed Mills

Not all mills had four sails—a good many managed in an emergency to work with two. The object of adding extra arms was to increase the effective area of the sails, and with it the efficiency of the machine. There were some mills built with five, six or even eight sails. The five-sailed mills were very efficient although the loss of one sweep rendered the mill idle until it was repaired. With one sail out of action the whole balance of the sails was lost. Nevertheless more mills were built with five sails than with six. The last working five-sailer is the tower mill at Alford, Lincs, and there is a five-sailer at Burgh-le-Marsh, Lincs.

Six sails provided a better selection of alternatives. If one sweep became unusable its opposite partner could be put out of use and the balance restored. Even three sails out of order did not necessarily mean that the mill had to stop work, as alternative sweeps could be employed to maintain the balance. A magnificent example in full working order is Trader Mill, Sibsey, Lincs. Towers at Coleby Heath and Peterborough displayed the same sail arrangement.

According to Mr Rex Wailes in *The English Windmill* there were but seven eight-sailers in England—at Diss, Norfolk; Eye, Cambs; Wisbech, Cambs; Much Hadham, Herts; Holbeach, Rasen and Heckington, Lincs. This last mill, the sole survivor, is preserved by Lincolnshire County Council.

The Annular Sail (The Wind Wheel)

The tall tower at Haverhill, West Suffolk, had a vast wheel instead of the conventional form of sail. Around the perimeter were ranged 120 shutters—each one five feet in length. Access to the shutters was provided by an unusually high gallery. This mill had a high domed cap and a fantail. The upper part of the wheel was over 80 feet above the ground, and it is not surprising to find that it could easily be seen from the neighbouring counties of Cambridgeshire and Essex. There was once another mill with this type of sail in that part of the world—it stood at Roxwell, Essex, and was dismantled in 1897. Wind wheels of this kind may still be found at work in the Mediterranean—on Majorca.

A much smaller version of the Haverhill mill was once erected on top of a Sussex barn at Angmering. It operated a water pump, a turnip chopper and a corn mill. It ceased work over 50 years ago and details of its features are sparse. There can be no doubt, however, about its wind wheel, which is recorded in R. Thurston Hopkins' booklet *Windmills* published many years ago in Sussex.

9. Grinding the grain

The path taken by the grain through the millstones is shown in the diagram on page 22. Millers made good use of gravity long before it was 'discovered' by Isaac Newton. Corn was first taken, via the sack hoist, to the top of the mill, where it was placed in the grain bin. A chute led to the hopper positioned above the millstones. Grain trickled from the bottom of the hopper on to the feed shoe which was methodically shaken by the rotating damsel or by the shaker on the crutch pole. As the upper runner stone revolved so a few grains at a time were fed into the eye to be ground and expelled around the stones' circumference. Then the meal fell into the meal spout and finally into the bin on the floor below. The effect of the grinding process was to make the emerging meal quite warm, and anyone feeling it for the first time usually expresses surprise at its temperature.

In their heyday most mills operated two types of stone. Barley was worked on Derbyshire Peak stones, which were hard and suited to this type of grain. Flour was usually processed on French burr stones, which were harder and better for finer grinding. These latter stones were not made in one piece, like the Derbyshire Peaks, but in sections cemented together and bound with iron bands.

Only the upper (runner) stone revolved and it did not come into contact with the bedstone below as some people think. The space between the surfaces of the stones was minute and carefully controlled to produce the best results.

Not all stones revolved in the same direction, and it was their mode of revolution that determined the pattern of the dressing on the grinding faces. Clockwise or anti-clockwide motion depended upon the way the sails revolved and the layout of the machinery. Millstones which revolved clockwise were said to be easier to dress and in the conventional post mill, with two pairs

21

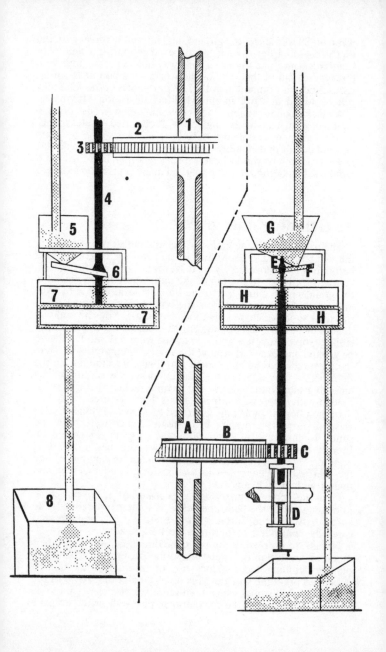

of stones in the head of the mill, this usually implied a counter-clockwise revolution of the sails. An observer can decide which way the sails revolved by noting the position of the sailstock or whip in relation to the sail frame: the sailstock always leads. Common sails are easier to ascertain if they are more or less intact.

10. Windmill facts and figures

From Primrose McConnell's *Agricultural Notebook* (1883)we may glimpse some of the technicalities associated with the operation of a windmill. He records the following data:

Angle of shaft of sails with horizon	=8° on level ground up to 15° on exposed heights.
Length of sails	=4 times the breadth.
Length of sails	=6/7ths of length of arm or 'whip'.
Arm divides sails	=proportion of 3 x 5; narrow part next the wind.
Area of sails	=¼ area of circle.
Area of sails	=⅞ of area of part occupied by vanes in small self-regulating windmills; this gives greatest effect.
Angle of sails to plane of motion	=5° at tip up to 22° next the axis.
Revolutions of millstone	=5 to 1 of sails.
Revolutions of sails	=12 per minute with the wind at a velocity of 20 feet per second.

A Windmill Archive has been established at the Museum of Lincolnshire Life, The Old Barracks, Burton Road, Lincoln LN1 3LY. This archive collects documentary records connected with windmills from all over Britain, as well as photographs, drawings and paintings. More information can be obtained from Mrs C. M. Wilson, Assistant Director (Museums and Windmills) at the Museum of Lincolnshire Life; telephone Lincoln (0522) 28448.

Opposite: **How the grain passes through the stones. Overdriven.** *1. Vertical shaft. 2. Great spur wheel. 3. Stone nut. 4. Crutch pole. 5. Hopper. 6. Feed shoe or slipper. 7. Millstones — the upper one rotates. 8. Meal bin.* **Underdriven.** *A. Vertical shaft. B. Great spur wheel. C. Stone nut. D. Jack ring — to lift stone nut out of gear. E. Damsel. F. Feed shoe or slipper. G. Hopper. H. Millstones — the lower or bedstone does not rotate. I. Meal bin.*

11. Some notable windmills

When the first edition of this guide appeared in 1968 the author thought he was chronicling the end of an era. To his delight events have proved him to be wrong. Today there are dozens of mills open to the public and many are in full working order. Private and public effort have made great advances in windmill restoration. Local authorities like Kent County Council have made systematic and long term efforts to care for the windmills in their areas. The outstanding work of Norfolk County Council, in conjunction with the Norfolk Windmills Trust, has resulted in the finest collection of restored windmills in England. Certain mills, some of great technical importance, have undergone years of voluntary effort. Without the determined dedication of those who have given up hundreds of hours of their spare time many projects would never have been completed. Most of the entries below are accompanied by telephone numbers. The visiting arrangements are correct at the time of going to press, but readers are advised to check details of opening times in advance of a visit. Where appropriate, arrangements for organised parties are also given.

The author will be pleased to receive corrections and notes on other mills for inclusion in a later edition. He is particularly interested in old mill photographs, documents concerning mills and in millers' recollections of working mills.

BEDFORDSHIRE
Stevington. post mill (1770). This is the finest mill in the county to have survived. It has two features worthy of note — a mansard roof and an unusually prominent flared skirt. This may have been the last working mill in the county with common sails. The mill, owned by Bedfordshire County Council, stands on the Bedford side of the village. Park at the bottom of Mill Lane and avoid the growing corn which usually surrounds the mill in summer. The key is obtainable from the Royal George, Silver Street, Stevington between 10 a.m. and dusk (or 7 p.m.). Telephone Oakley (023 02) 2184. Visitors pay a returnable deposit and sign the visitors' book.

BUCKINGHAMSHIRE
Bradwell Mill, New Bradwell, Milton Keynes. Tower mill (*c* 1816). Situated about a quarter of a mile south of the A422 at New Bradwell close to the old railway line and the Grand Union Canal, the mill is still being restored. Local limestone was used to construct the tower. As most Buckinghamshire tower mills were brick constructions this makes Bradwell Mill a particularly interesting example. Another unusual feature is its boat-shaped cap and the luffing gear which is operated by a chainwheel at the

rear. This gear also incorporates a universal joint — an early use of this device, which is now commonplace on motor vehicles. Originally the mill was worked with a pair of common sails and a pair of spring sails. There are two pairs of underdriven stones. Although the mill is relatively recent in date it did not employ the refinements of a fantail or patent sails. On the ground floor there is a fireplace — an extraordinary feature as flour dust is so potentially explosive.

Brill. Post mill (1668). This fine mill commands a view across the Vale of Oxford to the Cotswolds. Earlier in this century it had an open trestle, which is now protected by a brick roundhouse. A detailed drawing and description of the mill appears in Stanley Freese's *Windmills and Millwrighting*. This mill is opened by members of the Brill Society on Sundays from Easter to September. Arrangements can be made for educational groups on Mondays to Fridays through the Clerk to the Parish Council. Telephone Brill (0844) 237060.

Lacey Green. Smock mill (*c* 1650). Behind the Whip public house at the north end of the village, this mill is the oldest of its kind in England and its massive wooden machinery represents a significant part of our technological history. After many years of careful and painstaking restoration the task is now approaching completion. The work has been carried out by volunteers led by Christopher Wallis on behalf of the Chiltern Society, which has temporary custody of the structure. It is thought that the mill was moved here from Chesham in about 1821. All the basic machinery derives from the seventeenth century but the ancillary equipment is mostly Victorian. The mill is open May to September on Sunday and bank holiday afternoons, 3-6. Visitors should park on the public roadway and not drive up the short track leading to the mill. Group visits: telephone Princes Risborough (084 44) 3560.

Pitstone. Post mill (1627). This fine mill stands to the south side of Ivinghoe by the B488 to Tring. On the west side of this road (the right-hand side as you go towards Tring) there is room to park. The mill is clearly visible at this point, in the field beyond the stile. This is the oldest dated mill in England. The buck is typical of the post mills in this region. The tailpole is fitted with a wheel which made luffing easier. The machinery is intact, and a well illustrated guidebook is available. A model of this mill in its early nineteenth-century form can be seen at the Buckinghamshire County Museum, Church Street, Aylesbury. Pitstone mill is open on Sundays and bank holidays from May to September 2.30-6, or by special arrangement with I. A. Horn, telephone Cheddington (0296) 668227. School parties: Cheddington (0296) 668685. It is owned by the National Trust.

Quainton. Tower mill (1830). James Anstiss had the mill built and after 150 years it still remains in family ownership. The tower is 65 feet high and contains six floors. At the third floor level the gallery has been rebuilt. The domed cap and head frame assembly were recently replaced. Work on the sails is continuing. The iron brakewheel is of special interest as it is cast in a single piece. A steam engine formerly worked on the ground floor. There are several carved bricks near the base of the tower and two bear the date 1830. Restoration takes place on Sunday mornings and visitors and helpers are welcome. Visits by individuals and groups can be arranged with G. Rodwell, telephone Quainton (029 675) 348, or D. Moreton, Quainton (029 675) 224.

Wendover. Tower mill (1796). Wendover mill has the largest cap in the land. The hexagonal tower is 66 feet high and 25 feet across at the base, where the walls are 3 feet thick, and 18 feet wide at the curb. An ingenious winch worked by a worm gear once allowed the cap to be turned from the inside. The present cap was built by Derek Ogden. The mill is a private residence. No machinery survives. Visitors are not admitted but the mill may be observed from the A413.

CAMBRIDGESHIRE
Barnack. Tower mill. Most of the machinery is still in place but the sails have been removed. The mill ceased to work commercially in 1914. It is open on Saturdays and Sundays throughout the year and can also be viewed by appointment with the guardian, Mr K. I. Woolley, Windmill Farm, by telephoning the Burghley Estate Office. Telephone Stamford (0780) 52075. There is no admission charge.

Bourn. Post mill. Although Pitstone mill (Bucks) has the distinction of being the oldest dated mill it is not typical of the earliest post mills. Bourn reflects the pattern of the mills known to us from early carvings and manuscript illustrations. This mill has occupied its present site since 1636 and is probably the country's real veteran. It is open daily during the summer months. For access arrangements contact Cambridge Preservation Society, Wandlebury, Cambridge. Telephone Cambridge (0223) 243830.

Great Chishill. Post mill. This splendid open trestle mill on the west side of the village stands on an open site near two houses where the road starts to dip downhill. The carefully restored boarding is painted white. The spider testifies to the former presence of shutters. Above the mill steps the fantail stands upon its carriage, another feature unusual in this area. The interior may be inspected, at reasonable times, by arrangement with J. Pegram, Mill Bank, 72 Heydon Road, Great Chishill, near Royston, Herts SG9 8SR. Telephone Chrishall (0763) 838545. Parking is limited.

Great Gransden. Post mill (1674). This mill has some particularly fine machinery. It has been restored and it is eventually hoped to make the mill work again. The key can be obtained in the village — see notice board at the mill — or from Cambridgeshire County Council Property Department, telephone Cambridge (0223) 317323.

Histon. Smock mill. An attractive mill in a typical village setting, this has a high two-storey tarred octagonal brick base and two wooden storeys painted white. There is a fanstage and a cap gallery. It is private but can be viewed from the entrance to the driveway.

Madingley. Composite mill. This was formerly a post mill situated at Ellington and rebuilt in 1936 by C. J. Ison (of Histon). It stands on the site of a previous mill blown down in 1909, and has two common and two patent sails. It is privately owned and stands by the busy A45.

Over. Tower mill (*c* 1860). Less than a mile south of the village on the Longstanton road and next to the railway cutting, this four-storeyed left-handed tower mill incorporates parts of the earlier smock mill which stood on the site. The mill ceased work in 1929 and was restored to working order in 1968. The cap, fantail and two sails are new. There are two pairs of underdriven stones, grain cleaner, flour dresser and elevators for visitors to see. The mill is open for the sale of flour produced by the mill, Monday to Friday 9-6, Saturday 9-4, Sunday 10-12, all year round. Visits by individuals and adult parties by appointment. Telephone Swavesey (0954) 30742.

Soham (Downfield Mill). Tower mill (1726) working with two sails. The upper half of the tower has tapered (battered) sides, which give it an interesting profile. The windshaft of iron has a timber brakewheel. This is a working mill, in use throughout the year. Wholemeal flour and bran are always available. Visitors are admitted on Sundays and bank holidays (Christmas and New Year excepted) from 11 to 5. No admission charge for flour sales. Weekday visitors by arrangement with Nigel Moon, Leicester (0533) 707625. Local enquiries to Roger Allison, 23

CAMBRIDGESHIRE

Military Road, Soham. The mill is located at the south side of the village near the junction of the A142 and A1123.

Wicken Fen. Drainage mill. This four-sided smock mill was moved here from Adventurers Fen (part of Burwell Fen) and it was originally known as Norman's Mill. A plate records the details of its removal. It has four common sails and a small cap that can be turned with the Dutch-style tailpole. The door at the south side can be opened so that visitors can see the structure of the large scoop wheel within. It has been recently overhauled, and is now in full working order and can be used to help maintain a high water table on the nature reserve. Look for the National Trust sign as you enter Wicken village and then turn down Lode Lane to the Fen. There is a good car park, with toilets. The nature reserve on which the mill stands is National Trust property and is open daily except Christmas Day. Parties by arrangement only. Telephone Ely (0353) 720274.

DERBYSHIRE

Dale Abbey (Cat and Fiddle Windmill), near Ilkeston. Post mill. Built about the beginning of the nineteenth century, the mill was still working in 1952. Though in poor repair, it is complete and there are plans to restore it to full working order. Visiting is by prior appointment only, from April to the end of September. Contact the owners, Mr and Mrs Richardson, Dale Abbey Windmill, Cat and Fiddle Lane, Dale Abbey, Ilkeston. Telephone Ilkeston (0602) 301585.

Heage. Tower mill. Rebuilt after serious gale damage in 1893, Heage windmill worked until 1919. Since it was bought by Derbyshire County Council in 1966 it has been restored under the direction of millwright Mr R. Thompson and is the only complete example of its type in the county. Access is by appointment only through the County Planning Officer at Matlock. Telephone Matlock (0629) 3411, extension 7164. From Ambergate take the road to Nether Heage. At the Heage Miners' Welfare turn up the lane to the windmill on the hill.

DORSET

Cann Mills, Shaftesbury. Tower mill (1969). This typical Portuguese tower mill, built on top of a working watermill, is not open to the general public but a good view of the mill can be obtained from the Lower Blandford Road, A350. Some visitors are admitted by appointment with Norman Stoate. Telephone Shaftesbury (0747) 2475.

EAST SUSSEX

Argos Hill, Mayfield. Post mill (*c* 1835). This mill worked almost continuously until 1927. There are two pairs of over-driven stones and the usual ancillary machinery. With its red roof, white painted sails and distinctive fan carriage the mill is a significant landmark. At present, the mill is not open to the public.

Cross-in-Hand. Post mill. This mill was moved twice (in 1855 and 1868) before it reached its present site. It originally stood at Mount Ephraim, Uckfield. An unusual feature is the large metalled body. There is a fine fan carriage on the tailpole. The roundhouse has two storeys, of tarred brick and galvanised iron. The mill is not regularly open but special arrangements can be made with the owner, Mr Newnham, Hillside, Cross-in-Hand.

Nutley. Post mill (*c* 1670). This interesting open trestle mill has a tall slender body. It has not worked for about sixty years. Then it was operated with spring set sails. It has two pairs of stones, which are arranged with the burrs in the head and the Peaks in the tail. The mill may be reached from the Nutley to Crowborough road about a mile from Nutley. It is open on the last Sunday of each month from Easter to September, 2.30-5.30.

Polegate. Tower mill (1817). The mill stands in Park Croft off the A22 near Polegate cross roads, four miles north of Eastbourne. There is a museum area adjacent to the mill and other milling exhibits are in the mill itself. All the mill machinery is complete. There are three sets of stones. The brake wheel, on its cast iron windshaft, still has its applewood teeth. When the wind allows, the sails are turned but the stones are usually left out of gear. The striking rod and vane are complete and in working order. The fanwheel has five blades. The mill was worked by wind until 1943. After restoration the mill was opened to the public in 1967. A small car park adjoins the mill but visitors may also park on the public road. Open Sundays and bank holiday Mondays from Easter to September and Wednesdays in August, 2.30-5.30. Parties at other times by arrangement with C. E. Waite, 48a, Wannock Lane, Lower Willingdon, East Sussex. Telephone Polegate (032 12) 4763.

West Blatchington. Smock mill (*c* 1820). This is one of England's most unusual and attractive mills. The smock tower stands upon a square flint and brick base. A long range of barns formerly extended north and south from this structure. Gears for the underdriven stones are contained within the masonry tower, and the stone floor is at the level of the unfenced gallery. An auxiliary drive from the spur wheel provided power for the barn machinery, which once included a threshing drum. The smock tower has three floors below the cap, which was once turned by a six-bladed fan. The mill ceased work about 1900. Now restored, the mill is located in Holmes Avenue, Hove, north of the Old Shoreham Road (A27). It is open on Sundays and bank holidays from May to September, 2.30-5. Advanced bookings and special out of season visits may be arranged with Peter Hill, Brighton (0273) 734476, or Hove Planning Department. Brighton (0273) 775400.

ESSEX

Ashdon, near Saffron Walden. Post mill. Built in the mid eighteenth century, this small mill ceased to operate commercially early in the twentieth century. In recent years it has been restored but is not operational. Access is by appointment only with Major Bell, Manor Farm, Horseheath, Cambridge. Telephone Cambridge (0223) 891776.

Aythorpe Roding. Post mill (c 1760). An unusually large post mill with a body about 45 feet high. Restoration work by the County Council is now complete. Arrangements to visit the mill can be made with the County Planner, Essex County Council, Chelmsford. Telephone Chelmsford (0245) 352232, extension 293.

Bocking Churchstreet. Post mill (c 1680). This attractive mill with its tall roundhouse formerly stood on a site a quarter of a mile away. It was moved to its present location in 1830. The roundhouse contains an interesting collection of milling bygones. The roundhouse wall also served as an advertising space and it still proclaims the fact that grist was ground at the shortest notice. This means that the miller would grind the corn taken to him. The mill was restored in 1964 by Noble and Sons of Ongar. The key is held by Miss Tabor in the adjacent bungalow.

Finchingfield. Post mill (c 1760). The mill stands on the north side of the village beside the Hempstead road. It is under repair by the County Council and is not presently open to the public.

Mountnessing. Post mill (1807). Repairs to this very attractive landscape feature were completed in 1984. The mill is unusual as its roundhouse has sixteen sides. Frequently opened at weekends during summer, or at other times by arrangement. Details from Mr R. Wooding, 190 Rayleigh Road, Hutton, Brentwood, telephone Brentwood (0277) 215777, or from the County Planning Department, Essex County Council. Telephone Chelmsford (0245) 352232, extension 293.

Ramsey. Post mill (1842). This mill has several important technical features and is one of the most interesting in the county. Its three-storey roundhouse is unusual for Essex and its three-floored buck gives the mill a Suffolk appearance. The double shuttered sails were very up-to-date when the mill was built; they are mounted on a cast iron windshaft. Another rare detail was the fantail mounted upon the rear gable instead of upon a carriage above the steps. The mill worked by wind until 1939. The mill was restored between 1973 and 1978 entirely by voluntary labour under the direction of Christopher Hullcoop. Some thirty-six institutions, public bodies or companies made grants, provided materials or loaned equipment. The mill is not open to the public.

Rayleigh. Tower mill (1798). Located in Bellingham Lane, this mill has had a chequered past. Early in this century the cap and sails were removed and a steam engine installed. The mill continued to work until the 1920s. Although it has now lost all its machinery the sails and cap were restored in 1974. Its tower is a distinctive shape as the first two floors are vertical and those above have a pronounced batter (slope). The ground floor, the only part now open to the public, contains a permanent collection of photographs and other local exhibits. The mill is normally open on Saturdays 10-12.30, April to September, but visitors are also admitted at other times by prior appointment with Mr R. D. Miller, 33 Albany Road, Rayleigh, telephone Rayleigh (0268) 775768, or Mrs S. Green, 131 Daws Heath Road, Rayleigh, telephone Rayleigh (0268) 774897. Please enclose s.a.e. for a reply.

Stansted Mountfitchet. Tower mill (1787). The mill, scheduled as an ancient monument, was built by Joseph Linsell and remained in use until 1910. All the machinery remains. One of the three pairs of stones has been dismantled so that visitors can see their working faces. The mill now belongs to the people of Stansted and is open on the first Sunday in the month from April to October and every Sunday in August 2.30-7. On bank holidays the mill is open Sundays and Mondays. Arrangements for parties at other times. Telephone Bishop's Stortford (0279) 813159.

Stock. Tower mill (*c* 1800). This tall brick tower has a boat-shaped cap and two double shuttered and two single shuttered sails. The window openings are arranged spirally. Repairs are still being made by Essex County Council and for the present the mill cannot receive visitors.

Thaxted. Tower mill (1804). The mill stands at the edge of the town near the church. On the ground floor there is a rural museum which includes an 1835 fire engine and hand tools of many trades. The restored mill is an important landscape feature. There are three pairs of stones. Most of the machinery is intact but not yet in working order. Originally the mill had spring sails, like Outwood — see below, but it was fitted with patent sweeps about 1835. After a century of use it ceased work and began to decay. The restoration began with fund raising in 1970 and over the years most of the work has been completed. The mill is open May to September on Saturdays, Sundays and bank holidays, 2-6. Special arrangements for parties on application to Mark Arman, Hanna's, Bolford Street, Thaxted, Essex CM6 2PY. Telephone Thaxted (0371) 830366.

Tiptree (Messing Maypole Windmill). Situated in Church Road, this late eighteenth-century brick-built tower mill ceased work in 1936 and is now a private house. Some machinery remains and may be viewed by appointment with the owner, Ms A. V. Kinsella-Jaques. Telephone Tiptree (0621) 815187.

HEREFORD AND WORCESTER

Avoncroft Museum of Buildings, Stoke Heath, Bromsgrove. One of this important museum's most outstanding exhibits is the reconstructed post mill from Danzey Green, near Tanworth-in-Arden. Its clasp arm brakewheel drives the wallower mounted on the stone spindle of the overdriven stones. Four new common sails have been fitted and the sail span is just over 60 feet. When conditions allow, the mill is operated. Stone-ground flour is on sale in the museum shop. Visitors will find many other exhibits of interest, including a fifteenth-century hall house from Wales, and a 1946 prefab. The museum is open from 1st March to 30th November at the following times: June, July and August daily, 11 - 5.30; April, May, September and October daily except Mondays 11 - 5.30, open on bank holidays; March and November 11 - 4.30 but closed Mondays and Fridays, open on bank holidays. Telephone: Bromsgrove (0527) 31363.

HERTFORDSHIRE

Cromer. Post mill. The mill has an ogee-shaped gable. It is a typical white boarded East Anglian mill. During the past fifty years it has been repaired several times. It is not open to the public. The exterior only can be viewed. It stands close to the roadway on the Buntingford side of the village.

1. West Blatchington windmill, Hove, East Sussex.

2. Rottingdean smock mill, East Sussex.

3. White Mill, Sandwich, Kent.

4. Thaxted tower mill, Essex.

5. Pitstone post mill, Buckinghamshire.

6. Wilton tower mill, Wiltshire.

7. Wooden windpump from Pevensey, now at the Weald and Downland Open Air Museum, Singleton, West Sussex.

8. Thorpeness post mill, Suffolk.

9. Stracey Arms tower mill, Norfolk.

10. Dobson's Windmill, Burgh-le-Marsh, Lincolnshire.

11. Eastbridge Windpump, re-erected at the Museum of East Anglian Life, Stowmarket, Suffolk.

12. Wrawby post mill, near Brigg, South Humberside.

13. *Shipley smock mill, West Sussex.*

14. Trader Mill, Sibsey, Lincolnshire.
15. Hunsett windpump, Norfolk.

16. *Danzey Green post mill at Avoncroft Museum, Bromsgrove.*
17. *Ashton Windmill, Chapel Allerton, Somerset.*

18 (above). Great Chishill, Cambridgeshire. Tailstones, showing the tailwheel, wallower and hopper (right).

19 (left). Great Chishill, Cambridgeshire. Flour bin, chute, steps and (at top right) centrifugal governors to control the distance between the millstones.

HUMBERSIDE

Hessle. Tower mill. This is the only surviving example of a windmill used to work a whiting mill and is preserved by the Beverley Borough Council. The stone crushing tub and edge runner stones survive and have been refixed but it is unlikely the cap and sails will be replaced in the foreseeable future. The mill is open to the public on Sunday afternoons during the summer.

Skidby. Tower mill (1821). The mill, shown on the cover, was built in 1821 to replace an earlier post mill. It is preserved in full working order by the owners, Beverley Borough Council, the millwrighting work being carried out by R. Thompson and Sons of Alford, who have been maintaining the mill since 1913. The mill has four double patent sails and three pairs of stones, each 4 feet 6 inches in diameter, one Peak, one French and one composition. Wind power was used until 1954, when the mill was converted to electricity. The stones were left *in situ* but one of the adjoining warehouses was fitted with roller machinery recovered from a flour mill in Hull. The mill ceased operation in 1966, following which it was transferred to the Council. A museum of windmilling (including a reading room with research information on East Yorkshire windmills) and agriculture has been established in the adjoining range of buildings. There is also a working smithy and a wheelwright's workshop. The mill is open May to September, Tuesday to Sunday, and is operated on alternate Sunday afternoons commencing the first Sunday in May. From October to April the mill is open Monday to Friday. Party visits can be arranged by telephoning Mrs Woodcock on Hull (0482) 882255 and stone-ground flour and souvenirs can be purchased.

Waltham, Grimsby. Tower mill. This typical six-sailed mill now works with four sweeps. It is open from Easter to the end of September on Saturday and Sundays, 11-5. Annual open days are held at bank holidays in spring and summer, and these are accompanied by special attractions such as steam threshing and traction engines. Special arrangements can be made for parties on application to Wilf Wilson, 4 Paignton Court, Scartho, Grimsby DN33 3DH. Telephone Grimsby (0472) 825368.

Wrawby. Post mill (*c* 1790). This mill was carefully restored 1961-5 by a group of local enthusiasts with the help of Messrs Thompson of Alford. There are two pairs of stones, in head and tail. The French stones (head) are 4 feet 8 inches in diameter and the Peak stones 4 feet 2 inches. Coil springs control the sail blades, called shades in this part of the country, and the sails are mounted on an iron cross arm. There is a collection of milling bygones in the roundhouse. Wrawby mill ceased commercial work in 1940 when it lost two sweeps. In December 1979 the mill became tailwinded and the sails were destroyed. By Easter 1981 the mill was back in working order. Open on Easter Monday, May Day, Spring and August Bank Holiday Mondays, and on the last Sunday in June and July. Visits for individuals or groups can be arranged at other times on application to Richard Day. Telephone Brigg (0652) 53699.

ISLE OF WIGHT

Bembridge. Tower mill (*c* 1700). This fine tower mill was last used in 1913 and was restored 1959-61 and is in the care of the National Trust. Cars can be parked at the end of the bridleway leading to the mill. Open April to September daily 10-5 (last admission 4.45), closed Saturdays in April, May, June and September. School parties by appointment with the custodian, telephone Bembridge (0983) 873945, during the season. Out of season visits by arrangement with the National Trust, telephone Bembridge (0983) 526445.

KENT

Chillenden. Post mill (1868). Located on farm land directly north of the village. This fine mill has an open trestle. Open July to September on Sunday afternoons and bank holidays. Special arrangements: telephone Canterbury (0227) 720464 or Nonington (0304) 840329.

Chislet, near Reculver. Smock mill. Last worked in 1916, this is one of Kent's oldest surviving windmills, built in the mid eighteenth century. The sails have been removed but inside much of the machinery is of wood. Viewing is by appointment only with the owners Mr and Mrs A. Cook, The Old Millhouse. Telephone Chislet (022 786) 267.

Cranbrook (Union Mill). Smock mill (1814). The white octagonal tower of this glorious mill is 75 feet tall and the sweeps 68 feet across. Patent sails replaced the original common ones in 1840 when the fantail was added. The name is derived from the 'union' which was formed in 1819 when its first owner, Henry Dobell, became bankrupt. His name appears with the builder's on a stone tablet set in the brickwork. The mill was extensively repaired 1950-60. Vehicles may not be taken into the mill yard. There is a public car park at the bottom of Mill Hill 50 yards away. Open Easter to September on Saturday afternoons and bank holidays. Special arrangements: telephone Cranbrook (0580) 712256.

Eastry, near Sandwich. Smock mill. This mill is believed to date from the eighteenth century, the last of a series of six which have stood in this vicinity. Of the machinery only the windshaft, brakewheel and brake remain.

Herne. Smock mill. Mounted on a two-storey brick base, the mill was built in 1781 and ceased working by wind power in 1952. Until a few years ago milling continued, powered by electricity. The tower retains a timber stage from which to adjust the sails and one pair of patent sails. Much of the machinery is timber and there are two pairs of French burr stones and one pair of Derbyshire Peak stones. Open late July to September on Sunday afternoons and bank holidays. Special arrangements with Friends of Herne Mill, telephone Herne Bay (0227) 363345.

Margate (Draper's Mill). Smock mill (*c* 1845). The mill, in College Road, was restored to its working state in 1968-75. It has four floors. There are three sets of stones. Its white shuttered sails contrast with the black tower. The mill provides an important landmark. Grinding depends upon sufficient wind and the availability of a miller. There are good parking facilities outside the mill, which is open on Sunday afternoons, May to September. Special arrangements with Draper's Mill Trust, telephone Thanet (0843) 291696.

Meopham. Smock mill (1801). This hexagonal mill with five floors was largely restored in 1962 by E. Hole and Son of Burgess Hill. The original machinery was modernised in 1920 but the mill ceased to work by wind in 1929. New sails have now been fitted and the restoration of the machinery completed so that grinding can be resumed. The mill stands on A227 overlooking Meopham Green. Open Sunday afternoons and bank holidays, June to September. Special arrangements with Meopham Mill Trust, telephone Meopham (0474) 813218 or 812110.

Sandwich (White Mill). Smock mill (*c* 1830). This fine mill has an octagonal plan, a typical cap and unusual double shuttered sails controlled by half bow springs. The mill is now administered by the White Mill Folk Museum Trust. The miller's cottage is being restored as part of the Trust's programme. The mill is open every Sunday and bank holiday from Easter Sunday until the middle of September, 2.30-5.30. Special arrangements for parties can be made with Mrs Barber, telephone Sandwich (0304) 612076.

Stelling Minnis. Smock mill (1878). A tarred tower with a neat cap and a fanstage. Open Sunday afternoons and bank holidays from April to September. Special visits arranged by Stelling Minnis Parish Council, telephone Stelling Minnis (022 787) 291.

West Kingsdown. Smock mill. This black-towered mill contains three pairs of stones. The mill is located close to the A20 between Farningham and Wrotham. It was originally situated about a mile away but was moved to its present position in 1880. The Ordnance Survey map of 1805 shows its former position. Open at any reasonable time by arrangement with Mr D. Heaton, Mill House, West Kingsdown, Sevenoaks, Kent. Telephone: West Kingsdown (047 485) 2357.

Wittersham (Stocks Mill). Post mill. Erected here in 1781, although it may have been transferred from a previous site, the mill ceased to work in 1900. It was then used as a dwelling and much of the machinery was removed. Since it was given to the Kent County Council in 1980 the structure has been extensively restored. Open Sunday afternoons and bank holidays, May to September. Special arrangements with Friends of Stocks Mill, telephone Wittersham (079 77) 309.

Woodchurch. Smock mill. Built about 1820, the mill was worked until 1926 but, although intact, was near to collapse when restoration was begun. A new tower was erected in 1981 and new floors are in place. While restoration continues, the machinery is open on Sundays, 2-4, from Easter to the end of September and at other times by appointment. Telephone Woodchurch (023 386) 519.

LANCASHIRE

Thornton (Marsh Mill). Tower mill (1794). The 70 foot tall tower contains the original machinery. The mill has been carefully restored and is open to visitors during June, July and August on Sundays from 2 to 6 p.m. and on Wednesdays from 6 to 9 p.m. Parties are admitted at other times by arrangement with Mr Walter Heapy, telephone Cleveleys (0253) 826295, or Wyre Borough Council, Marine Hall, Fleetwood, Blackpool; telephone Fleetwood (039 17) 71141.

LEICESTERSHIRE

Kibworth Harcourt (9 miles south-east of Leicester). Post mill. Built in the middle of the seventeenth century and last worked in 1920, the mill has been completely restored, with all the machinery intact. It is owned by the Society for the Protection of Ancient Buildings and may be viewed by appointment with Mr B. J. E. Briggs, Windmill Farm. Telephone Kibworth (053 753) 2413.

Wymondham, near Melton Mowbray. Tower mill. When it was built in 1814 the five-storeyed mill had six sails. It was wind-powered until 1918, when a steam engine was installed which operated until 1952. Most of the original machinery remains and it is undergoing complete restoration, after which it will be open to the public. Enquiries to the owner, Mr W. H. Naylor, The Mill House. Telephone Wymondham (057 284) 639.

LINCOLNSHIRE

Alford. Tower mill (1813). Alford is a five-sailed mill which is still worked. Standing 75 feet tall with its white sails and tarred tower, this is one of England's outstanding windmills. There are six floors below the elegant ogee cap. The iron gearing was made by Tuxford of Boston and the firm's name appears on the great spur wheel. Mr Banks, who works the mill, comes from an old milling family. His father was the miller at Sturton by Stow for thirty years. This mill has four pairs of stones on the stone floor (two pairs of Derbyshire greystones and two pairs of French burr stones) and one pair of greystones in the bottom floor of the mill to be driven from outside by tractor or other power. A guide leaflet, postcards and stone ground floor are on sale. Car parking is limited, but a lay-by down the Louth Road is suitable for coaches. Opening hours not at present fixed. Enquiries to Catherine M. Wilson, Lincolnshire County Council, telephone Lincoln (0522) 29931, extension 2805.

Burgh-le-Marsh. Tower mill (1833). The mill was purchased by the county council in 1965 and has been carefully restored. Its slender tarred brick tower is crowned with a typical Lincolnshire white cap. The five patent sails rotate clockwise. The tower has five floors. The third floor contains the wind driven stones and the visitor can examine in detail the grinding process. The stones on the first floor were worked by a diesel engine housed outside the tower. The mill is situated on A158, the Lincoln-Skegness road, and is on the Skegness side of the village. Open and working (weather permitting) on the second and last Sunday in each month 1-4. Otherwise open daily 10-5.30. Telephone Skegness (0754) 810281.

Heckington. Tower mill (1830). This is England's last eight-sailer. Its present cap and sails were brought here from Boston in 1892 by John Pocklington. The mill was originally built as a five-sailer. Its replacement cap was a little too wide for the tower and visitors can still see the difference. It formerly worked five pairs of stones and a saw mill. From the gallery there are good views across the surrounding landscape. The mill stands at the southern end of the village close to the railway station. It dominates the skyline and is one of the most spectacular in the country. It is now in full working order. Open Easter to September, Saturdays, Sundays and bank holidays, 2-4.30. Sundays only in winter. Further information from the adjacent Craft Centre, telephone Heckington (0529) 60765.

LINCOLNSHIRE

Lincoln (Ellis' Mill, Mills Road). Tower mill. Built in 1798, this was once one of nine mills in the vicinity. Frank Ellis was the last miller, until the 1940s when all the machinery was removed. Restoration began in 1977 when the Lincoln Civic Trust acquired it. Cap gear for the then derelict mill was acquired from Subscription Mill, Stourton-by-Stow, and stones from Enos' Mill, Toynton All Saints. The millwrights Thompson of Alford provided the sails and fantail. In 1981 it opened in full working order. Open October to April, second and fourth weekends in the month 2-6. May to September each weekend, 2-6. Special visits can be arranged with the Honorary Miller, Barry Brooke, 30 Harpswell Road, Lincoln (please provide s.a.e.). There is a small car park.

Sibsey (Trader Mill). Tower mill (1877). This magnificent mill has a 74 foot brick tower. It was built by John Saunderson of Louth. It has six sails. Now restored and owned by the Historic Buildings and Monuments Commission (English Heritage), it is open from April to September, weekdays 9.30-6.30, Sundays 2-6.30.

LONDON

Brixton Mill, Blenheim Gardens, Lambeth. Tower mill. This tarred brick mill is 48 feet with a boat-shaped cap. The mill worked from 1816 until 1934. In 1957 the mill was bought by the LCC. In 1964 it was restored and some of its present machinery was obtained from a defunct mill at Burgh-le-Marsh, Lincolnshire. The new sails were made by Thompson and Son of Alford, Lincolnshire. The mill is now owned by the Borough of Lambeth and is open to the public. Details from Lambeth Amenity Services, telephone 01-622 6655.

Shirley Mill, Upper Shirley Road, Croydon. Smock mill (*c* 1855). Located in the grounds of John Ruskin High School, the mill can at present be viewed only from the exterior by arrangement with the headmaster and/or the school keeper. Telephone 01-656 0994. This smock mill is an impressive example with four storeys and cap. There are two pairs of stones on the second floor. Eventually it is hoped to establish an agricultural museum on this site.

Upminster. Smock mill (*c* 1800). This impressive mill, with its tall white painted octagonal tower, stands in St Mary's Lane. One of the mill's interesting features is its boat-shaped cap which has a cap gallery in the Norfolk fashion. The sails are double shuttered. Upminster was formerly part of Essex and the mill was acquired by the Essex County Council in 1937. After repairs the mill was opened to the public in 1967. A scale model of it may be seen in Romford Central Library. For details of open weekends during the summer, contact the Borough Librarian, Central Library, St Edwards Way, Romford. Telephone Romford (0708) 44298 or 46040 extension 355.

Wimbledon Common, SW19. Hollow-post mill (1817). This attractive mill is the sole surviving hollow-post flour mill in England. Another similar mill once stood in Southwark near the site of the old Globe Theatre. One miller, Thomas Dann, also served as a constable and had the obligation to apprehend duellists who found the remoteness of the place suited to their needs! A more peaceful pursuit was followed by Robert Baden-Powell, who began to write his famous *Scouting for Boys* in the miller's cottage (1907). In 1864 the hollow-post and other principal machinery were removed. When the mill was restored in 1978 the first floor of the roundhouse was made into a milling museum and, in addition to mill machinery, the visitor can admire some finely detailed models. An excellent guidebook and drawing of the mill in its working condition are also available. The mill is open Saturdays, Sundays and public holidays from April to October, 2-5. A large car park adjoins the mill. Parties by prior arrangement: telephone 01-788 7655. Conducted visits can also be arranged with the Honorary Curator: telephone 01-947 2825.

NORFOLK

Berney Arms. Windpump, tower mill. In a very isolated position beside the river Yare near Breydon Water, the mill is best approached by river or by train at Berney Arms station. On foot the way is long and difficult. With its seven storeys this is the tallest windpump in the country. Apart from lifting water it was also used for grinding cement clinker. The scoop wheel is particularly large. English Heritage has a permanent exhibition inside the mill, which is generally open to the public from April to September daily, 9.30-6.30. Readers are advised to check opening times before visiting. Telephone Great Yarmouth (0493) 700605.

Billingford. Tower mill. Located on the edge of Billingford Common, on A143 about a mile east of Scole, this attractive red brick mill, with its white boat-shaped cap and fantail, contains most of its original machinery. The Norfolk Windmills Trust opens it at weekends throughout the summer. At other times the key is available — see the notice on the mill door. Special arrangements can be made for parties. Telephone Norwich (0603) 611122, extension 5224.

Cley-next-the-Sea. Tower Mill (1713). This must be one of the best known mills in Britain, a favourite subject for artists and photographers. It was last worked in 1919 and now provides accommodation for tourists. The mill is open on public holidays. Telephone Cley (0263) 740209.

Denver. Tower mill (1835). Located on the minor road to Denver Sluice from Denver village, this very attractive mill has most of its machinery, and the granary is now a well arranged museum. All the exhibits relate to the mill. Visitors can also see the steam mill which was latterly powered by a 45 hp Blackstone oil engine *(in situ)*. Open throughout the year excluding Sundays, 9 to dusk. Special arrangements for parties. Telephone Mr Staines, Downham Market (036 63) 2285; or Mr Chapman, Downham Market (036 63) 2188.

Garboldisham, near Diss. Post mill. Now the only complete post mill in Norfolk, it was built between 1770 and 1780 and ceased to operate in 1917. By 1972 it was in an advanced state of decay and the new owners began extensive restorations. There are two pairs of millstones in the head and mainly cast iron machinery from around 1830. The mill is open on weekdays 9-1, not bank holidays. Enquiries to Adrian Colman, Garboldisham (095 381) 593.

Great Bircham. Tower mill (1846). The mill stands on high ground to the west of the village, about three miles south of Docking and to the west of B1153. It has four sails and a six bladed fan winding the ogee cap. From the stone floor gallery there are fine views over the surrounding countryside. The machinery is in working order and the sails turn when there is sufficient wind. A bakery museum adjoins the mill. Good parking. Open Easter to late May, Sundays and Wednesdays, late May to end September daily except Saturdays, 10-6. Special arrangements telephone Syderstone (048 523) 393.

Horning (Hobbs Mill). Open-framed, timber trestle windpump. On the east bank of the river Bure, close to Wilds boatyard, this windpump is based on the traditional Norfolk tower mill design and drives a scoopwheel. Since the Norfolk Windmills Trust moved to save the windpump from collapse in 1981, its complete restoration is planned with the help of the boatyard staff and finance from Wilds and the Broads Authority. The windpump can be seen from the river but can only be reached on foot through the boatyard. Anyone wishing to visit should contact Wilds first.

Horsey. Windpump, tower mill. Located by B1159 at Horsey, about three miles north of Winterton. A National Trust property. Open mid April to end of September daily 11-5 (July and August 11-6). Gallery closed May, June, and possibly part of April and July.

How Hill (Boardman's Mill). Open-framed, timber trestle windpump. On the east bank of the river Ant at the How Hill Nature Centre, near Ludham, the mill is the only trestle mill left with a turbine and has a miniature cap, sails and fantail in the traditional Norfolk tower mill style, restored in 1981. It is owned by the Norfolk Windmills Trust, which has restored it with the help of a grant from the Broads Authority. The riverside footpath passes close by and the mill is open all year.

How Hill (Turf Fen Windpump). Tower windpump. Under restoration by the Norfolk Windmills Trust. On the west bank of the river Ant at the How Hill Nature Centre, near Ludham, this mill is clearly visible from the public path on the other side of the river. It is the only Broadland windpump where water levels are the same as they were at the time of building.

Little Cressingham (2 miles west of Walton). Wind and water mill. In this unique mill two pairs of stones on the first floor were driven by power from the waterwheel and two on the third floor were driven by wind. As well as the milling machinery, there is a Bramah pump and a hydraulic ram, both of which were used to supply water to the gardens of Clermont Lodge. Since 1981 it has been undergoing restoration by the Norfolk Windmills Trust. The mill is open to the public on open days, or by appointment with the Technical Adviser to the Norfolk Windmills Trust. Telephone Norwich (0603) 611122, extension 5224.

St Olaves. Windpump, smock mill. Located on the east bank of the river Waveney below St Olaves bridge on A143, the mill is visible from road or river and can be approached by the footpath leading from the bridge. This interesting tower is under restoration and in the care of the Norfolk Windmills Trust.

Starston. Hollow post mill (*c* 1850). At the western end of Starston village, this is a simple hollow-post drainage mill with four white spring set sails and two tail fins. Although of interest to the technical enthusiast, it is not as visually attractive as most other mills in the county.

Stow Mill, Paston. Tower mill (1827). Located by the B1159 coast road half a mile south of Mundesley, this is an attractive mill with a restored cap and fantail. It last worked some fifty years ago, when it was converted into a dwelling. It is now being restored to a working state, using machinery salvaged from other mills. A cardboard cut-out model of the mill is available. Open April to September, weekdays 2 to dusk, weekends 10 to dusk. Parties by arrangement with Mr and Mrs Newton, Paston Mill Cottage, Paston. Telephone Mundesley (0263) 720298.

Stracey Arms. Windpump, tower mill. Prominently positioned between the south bank of the river Bure and the A47, Acle to Great Yarmouth road, about three miles east of Acle, this four storey brick tower has a white boat-shaped cap and fanstage. It is open at Easter and from May to September daily, 9-8. Park at the adjacent restaurant and walk along the riverbank to the mill.

Sutton, near Stalham. Tower mill (1789) and Broads Museum. The mill stands about a mile and a half south-east of the junction of B1151 and A149 at Stalham and is signposted from Sutton village. This is England's tallest windmill. It has nine floors. There is an external gallery at the fifth floor level. The basic machinery is intact. A new cap and working fan wheel (12 feet in diameter) have been built and the resident owner intends to complete the restoration. The mill was rebuilt in 1857 after a bad fire. It was last worked in 1940, when lightning struck the sails and set them alight. The mill has a boat-shaped cap in the Norfolk style. From the fanstage there is a magnificent view over the surrounding countryside. The museum, based on the Nunn family collection, illustrates craft tools. There is a large collection of domestic bygones. Follow the road signs on A149 in Sutton village. Open April to mid May, Sunday to Wednesday, mid May to end September, daily except Saturday, July and August daily 10-6. Arrangements for parties with C. Nunn, telephone Stalham (0692) 81195.

Thurne Dyke. Windpump, tower mill. Next to the river Thurne at the entrance to Thurne Dyke, west of the village, this distinctive white tower with its pronounced batter has been a photographer's delight for many years. It was originally restored by its present owner, R. D. Morse, but is now leased to the Norfolk Windmills Trust, which maintains it with the generous financial aid of Messrs Hoseasons. For opening arrangements contact the Technical Adviser, Norfolk Windmills Trust. Telephone Norwich (0603) 611122, extension 5224.

Wymondham (Wicklewood). Tower mill. Built in about 1846, the five-storey mill ceased working only in 1942. It was given to the Norfolk Windmills Trust in 1977 and the work of restoration began immediately. The fabric has been restored and the machinery is being restored to full working order. Viewing is on open days or by appointment with Mr Woodrow, telephone Wymondham (0953) 603694.

NOTTINGHAMSHIRE

North Leverton (Subscription Mill). Tower mill (1813). This tower mill, like Cranbrook, was built by shareholders — hence its name. With its ogee cap, fantail and double shuttered patent sails, it shows its debt to the nearby mills of Lincolnshire. It has sloping sides up to the level of the third floor, where the tower becomes cylindrical. In 1959 the mill was struck by lightning and lost one of its sails. It is still worked commercially on behalf of its joint owners. Visitors are admitted on application if the mill is at work.

Nottingham (Green's Mill), Belvoir Hill, Sneiton. Tower mill (1807). This fine tower mill has been restored to full working order by the George Green Memorial Fund and Nottingham Museums. The mill was owned by George Green (1793 - 1841) who was not only a miller but also one of the outstanding scientists of his time. He devised new mathematical techniques which led to a greater understanding of magnetism, electricity and light. Green's Functions are still used by scientists in the fields of electronics, sub-atomic physics, telecommunications and engineering. He went to Cambridge University at the age of forty and died eight years later. He is buried in St Stephen's churchyard, close to the mill. The five-storey brick mill has been restored by R. Thompson and Sons. It has an ogee cap with fantail, two common sails and two spring sails. There are two pairs of overdriven French stones and a gallery at first-floor level. Around the mill yard is an educational Science Centre with working exhibits relating to Green's work, science, technology, and the history of milling and Green's Mill. The mill can be seen working whenever conditions allow. Open all the year round, Wednesday to Sunday and bank holidays, 10-12 and 1-5. For further information telephone the Science Centre; Nottingham (0602) 503635, or Nottingham City Leisure Services, Nottingham (0602) 411881.

OXFORDSHIRE

Bloxham. Post mill (1842). Situated at the end of the minor road that leaves the east side of A361 about half a mile north of Bloxham, next to the Warriner School, this mill resembles one in a medieval manuscript. Its open trestle, diminutive size and four common sails (restored in 1984) provide us with a view that must have been commonplace in feudal England. Although surrounded by growing crops in summer the mill can be viewed from the roadside. Bloxham mill is now in working order.

Wheatley Mill, Wheatley. Tower mill. This octagonal eighteenth-century mill stands on a trackway about a mile south-west of the church and at the top of the hill road leading to Garsington. It is being restored by the Wheatley Windmill Restoration Society and is not yet open at regular times. Visitors are admitted by appointment. Telephone Mrs M. Ramsden, Wheatley (086 77) 4610. Open days are advertised in the local press. Casual visitors are welcome on Sunday mornings when volunteers are at work. There is limited parking in Windmill Lane. Postcards are on sale at Windmill Cottage.

SOMERSET
High Ham (Stembridge Mill). Tower mill (*c* 1820). This west country mill is England's only surviving stone and thatched example. It is owned by the National Trust and is in a delightful setting next to the mill house. Open April to September on Sundays and bank holidays 2-5.30, or at other times by appointment. Special arrangements for parties out of season. Telephone Langport (0458) 250818.

SUFFOLK
Eastbridge Windpump. Re-erected at the Museum of East Anglian Life, Stowmarket. Smock windpump. This was one of four windpumps draining Minsmere Level near Leiston in Suffolk. Constructed in the late nineteenth century, it is, outwardly, exactly similar to a smock type of corn windmill with fantail winding and four patent sails. Inside, cast iron gearing on the upright shaft operates a three-throw crankshaft and a three-chamber lift pump which can raise water about 6 feet. This interesting pump had collapsed into the marshes after many years without maintenance. It was rescued by members of the Suffolk Mills Group and is again in working order — wind permitting. The museum, which includes a village store, blacksmith's forge, wheelwright's shop and domestic display, holds East Anglia's most important folklife collection. The Boby Building, a nineteenth-century woodworking and machinery erecting shop, contains major exhibitions of East Anglia's industrial heritage. There are demonstrations and craft events throughout the season. Parking on a large adjacent car park. Disabled drivers may drive to the mill and the watermill, if prior notice is given. The museum is open April to October, 11-5 Monday to Saturday and 12-5 Sunday (until 6 on Sundays from June to August). Party visits by arrangement. An education

centre is available for schools. Telephone Stowmarket (0449) 612229.

Framsden. Post mill (c 1760). This mill once had an open trestle. In 1836 the Suffolk millwright John Whitmore, of Wickham Market, modernised the mill by jacking it up 18 feet, building a roundhouse, a fan carriage and providing patent sails. The mill now operates occasionally on two sails. An adjoining stable contains a museum of country bygones. Since 1966 the mill has slowly been refurbished. A detailed account of the restoration appears in Brian Flint's *Suffolk Windmills*. Open on bank holiday Sundays, 11-6, and on some summer weekends. Car parking in the meadow next to the mill. For details write, with s.a.e., to John Ablett, Old Mill House, Framsden, Stowmarket. Telephone Helmingham (047 339) 328.

Herringfleet. Drainage smock mill (1820). The mill, on the river Waveney, can be reached by a public footpath from B1074. It has four common sails and its cap is turned by a tailpole of the Dutch type. The scoopwheel is 16 feet in diameter. Herringfleet mill continued to work until the 1950s. Open days are held several times a year, including New Year's Day. On these occasions the mill is put to work (wind permitting). Information from the County Planning Officer, St Edmund House, Rope Walk, Ipswich, Suffolk, telephone Ipswich (0473) 55801.

Holton Mill, Holton, Halesworth. Post mill (1749). This mill is preserved as a landmark. There is no internal machinery except the windshaft and brakewheel, but the fantail is in working order. Open Spring and August Bank Holiday Mondays. At other times by arrangement with the owner, J. Nichols. Telephone Halesworth (098 67) 2367.

Pakenham Mill, Pakenham. Tower mill. This is a fine example of a tower which is maintained in working order. The earlier cap was covered in copper but the new cladding is aluminium. The tarred tower is crowned by a domed cap that finishes in a traditional finial. From the fanstage a cap gallery extends part way around the circumference. The patent sails each have two rows of shutters. Open to individuals and parties by appointment only with M. Bryant. Telephone Pakenham (0359) 30277. Limited parking.

Saxtead Green, near Framlingham. Post mill (1854). There has probably been a mill on this site since the thirteenth century. In 1854 the structure was rebuilt and two extra floors were added to the roundhouse. With its patent sails and majestic fan carriage this is one of the most spectacular mills in the English countryside. It was in daily commercial use until 1947. Corn grinding no longer takes place, but when there is sufficient wind the sails are allowed to turn. The mill is in the care of English Heritage. Open April to September, Monday to Saturday 9.30-6.30. Closed 1-2. Not open Sundays. Telephone Earl Soham (072 882) 346 or Framlingham (0728) 82789. Parking as directed by the signs. Pedestrian access only to the mill.

Thelnetham. Tower mill (1819). This mill, notable for its dated cast iron windshaft (*J. AICKMAN. LYNN. 1832*), is being restored by the Suffolk Mills Group. The mill ceased work in 1924, and the restoration began in 1980. Its beehive cap has been renewed. When the four patent sails are replaced they will have a width of 9 feet 6 inches and a span of 64 feet. The mill is open on Easter Sunday, National Mills Day (the first or second Sunday in May), and August Bank Holiday Monday, 10-5. Parties and other visitors by arrangement with Mark Barnard, 41 Melbourne Road, Ipswich. Telephone Ipswich (0473) 77853. Car parking in adjacent paddock.

Thorpeness. Post mill (1824). Now used as an information centre for the Heritage Coast, this post mill was moved from Aldringham *c* 1923 when the milling gear was removed and pumping machinery installed. It worked as a pumping mill until 1939. The sails and fantail were renewed 1976-77. There is a large car park in the village. Open bank holiday weekends in May, June and September; daily (except Monday) during July and August, 2-5. Information from the County Planning Department, Suffolk County Council, St Edmund House, Rope Walk, Ipswich, telephone Ipswich (0473) 55801.

Woodbridge (Buttrum's Mill). Tower mill (*c* 1835). Open all the year round by arrangement with the Planning Department, Suffolk County Council. Telephone Ipswich (0473) 55801. The machinery is complete, but not in working order. There are four pairs of underdriven stones and an auxiliary steam drive.

SURREY

Outwood Common, Redhill. Post mill (1665). Located on the Common at the eastern end of the village and south of Harewood House parkland, this is the oldest working mill in the country. It has spring sails controlled by elliptical springs. It was restored in 1952. According to tradition men watched the Great Fire of London from the top of the mill in the year following its erection. Stone ground flour and a guidebook are on sale. There is car parking around the Common. The mill received a Civic Trust Award in 1980 and a Surrey Industrial History Group Award in 1984. It is open every Sunday, 2-6, from Easter to the end of October. Special arrangements can be made for parties and evening tours. Write to R. Thomas, Old Mill, Outwood Common, Redhill, Surrey, enclosing s.a.e. Telephone Smallfield (034 284) 3458.

TYNE AND WEAR

Newcastle upon Tyne (Chimney Windmill, Claremont Road). Smock mill. Built in 1782 and designed by John Smeaton, the mill worked until 1891. It was a golf club house until 1975 when it was bought by the architect Thomas Falconer who began a restoration and repair programme. The mill is now owned by Nigel Cabourn Limited. Visitors are welcome by appointment with N. J. Cabourn, telephone Tyneside (091) 232 3772.

Sunderland (Fulwell Mill). Tower mill (1821). This is one of Sunderland's best known landmarks and is currently under restoration. Virtually all the interior workings are intact. The mill is unusual in having a stone reefing stage, from which the sails could be adjusted. Details from Tyne and Wear Industrial Monuments Trust, Sandyford House, Archbold Terrace, Newcastle upon Tyne NE2 1ED. Telephone Newcastle upon Tyne (0632) 816144, extension 291.

WARWICKSHIRE

Chesterton. Tower mill (1632). Some writers have suggested that the tower was built as an observatory, and the design has been attributed to Inigo Jones. Documentary evidence, however, suggests that it functioned as a mill as early as 1647. The tower is about 36 feet high and has a diameter of 21 feet. Access to the mill interior is by way of a ladder. In its working days there was a staircase. Casual visitors have to be content with an exterior inspection. The anticlockwise sails have a span of 60 feet and carry 450 square feet of canvas. The machinery is unusual as most of the gearing is wooden. An 8 foot diameter brakewheel drives a lantern pinion wallower — in the Dutch style. There are two floors above the open arches, and the stones are on the lower one. They are placed upon a low hurst frame and are underdriven like the stones in most post mills. The shallow domed cap is rotated by an internal winch. This unique landscape feature was restored between 1965 and 1971. The mill is open to the public, usually during an autumn weekend, every two years. Car parking is then usually available in a nearby field. Open days are advertised locally and details are obtainable from the Department of Land and Buildings, PO Box 46, Shire Hall, Warwick CV34 4RP. Telephone Warwick (0926) 493431, extension 2435.

WEST MIDLANDS

Berkswell Mill, Balsall Common. Tower mill (1826). The mill is situated a quarter of a mile east of A452 and south of Balsall Common. It has a fine brick tower crowned by a well proportioned boat-shaped cap, which is turned by an endless chain. There are two common and two spring set clockwise sails, which have a span of 60 feet. The spring sales were operated by coil springs. A post mill formerly stood on this site and its cross trees are incorporated into the present mill. Old deeds show that there was a mill here in 1706 but the post mill could have been much older. All the machinery is intact. There are two pairs of underdriven stones — Peak and French burr. The mill ceased to work by wind in 1933 and a stationary engine was installed. Corn grinding continued until 1948. The restoration work, which began in 1973, was carried out by Derek Ogden. A descriptive leaflet, scale drawing and sketches of the mill are on sale.

Open May to September, Sundays only 10.30-12.30, 2.30-5.30, and at other times for individuals, schools or societies by appointment. Telephone Berkswell (0676) 33403.

WEST SUSSEX

Halnaker Mill, Halnaker Hill. Tower (c 1740). Situated north of A285 about a mile from Halnaker village, the mill commands fine views from within the ramparts of the fort, set at the top of Halnaker Hill. The sails are fixed. There is no internal machinery. Open to view at any time for those who trek the half mile up from the main road. West Sussex County Council.

King's Mill, Shipley. Smock mill (1879). This is the only remaining working smock mill in West Sussex. From 1906 to 1953 the mill was owned by the author Hilaire Belloc, who lived there and kept the mill working. The mill was restored in 1957 as a memorial to him. There is a permanent Belloc exhibition. King's Mill is still owned by the author's great grandson, Mr C. Eustace. Grinding takes place when the wind allows. There is much to see in this, the largest of the county's mills. The double shuttered sails are particularly impressive. Guidebook and postcards available. The Friends of Shipley Windmill provide refreshments for visitors on open days in the Shipley Village Hall. Arrangements can be made for party teas. There is roadside parking at the mill. Open on the first weekend of each month from May to October and on Spring and August Bank Holiday Mondays. Special arrangements can be made for parties on written application to Mrs A. Crowther, 13 Church Close, Shipley, Horsham, enclosing s.a.e. Telephone Coolham (040 387) 310.

Weald and Downland Open Air Museum, Singleton. The Museum is situated on A286 (Chichester to Midhurst road) about seven miles north of Chichester, and is clearly signposted. Among the re-erected buildings here is Pevensey windpump. These old wooden structures were largely replaced late in the last century when the more familiar steel alternatives were introduced from America. The Pevensey mill came from a clay pit. It has common sails. The gearing is of cast iron. It is an interesting technical survival which is fully described in the museum's excellent guide. Facilities for the disabled. Open April to October, daily 11-5; November to March. Wednesdays and Sundays only 11-4. Last admission one hour before closing. Parties and school visits by appointment. Telephone Singleton (024 363) 348.

WILTSHIRE

Wilton Mill, Wilton, near Great Bedwyn. Tower mill (1821). Half a mile north-east of Wilton village and south of the minor road that leads to Marten, this impressive mill with its tall domed cap is a lone survivor in this important corn growing region. It commands a fine view across the Vale of Pewsey. It has four floors above the ground. The two pairs of stones are under-driven. There are two common and two patent sails which have a spread of 64 feet. The cap is turned by a fantail that operates a worm wheel working on an external rack at the top of the tower wall. The mill is worked when qualified volunteers and sufficient wind are simultaneously available. There is parking on the adjoining highway verge. Open Sundays and bank holidays from Easter until the end of September, 2-5. Special arrangements for parties, in season, on application to Colonel A. D. Lewis, Midden Hollow, Wilton, Marlborough, Wiltshire. Telephone Marlborough (0672) 870268.

Chiseldon Mill, Windmill Hill Business Park, Swindon. The reconstructed mill tower has been erected here as a landscape feature. The new shuttered sails and fanstage are in working order but there is no internal machinery.

NORTHERN IRELAND

Ballycopeland Mill, Millisle, County Down. Tower mill (*c* 1784). This is the only working windmill in Northern Ireland. It was last worked commercially in 1915. There are four floors — hopper, stone, gear and ground. The tower is 33 feet high to the curb and the walls are 2 feet thick. The boat-shaped cap is similar to some mill caps in the north-west of England. Open April to September, Tuesday to Saturdays 10-1, 1.30-7, Sundays 2-7; October to March, Saturdays 10-1, 1.30-4, Sundays 2-4. Arrangements for parties. The mill operates by prior arrangement and if there is sufficient wind. Applications to Department of the Environment, Historic Monuments and Buildings Branch, Calbert House, 23 Castle Place, Belfast BT1 1FY, telephone Belfast (0232) 230560.

WALES

National Centre for Alternative Technology, Llwyngwern Quarry, Machynlleth, Powys. The centre is three miles north of Machynlleth just off A487 Machynlleth to Dolgellau road. There is a train service at Machynlleth and a bus service from Dolgellau and Machynlleth to Pantperthog (adjacent to the centre). Coaches should approach from Machynlleth. This fascinating project set on the fringe of the Snowdonia National Park demonstrates the possibilities of living with only a small share of the earth's resources. Examples of wind and water power include windpumps, aero-generators and an attractive Cretan-style mill with jib sails. Residential courses on all aspects of alternative technology. Send s.a.e. for details. Large bookshop. Refreshments available. There is a free car and coach park 300 yards from the Centre. Elderly and disabled visitors may drive up to the Centre, which is seventy feet up the hill. Open every day 10-5 (dusk in winter) except Christmas. Telephone Machynlleth (0654) 2400.

12. Gazetteer

The following lists include mills which are substantially intact. Many more remain in a partially dismantled state. Mills of this kind are recorded by Arthur Smith, Wilfred Seaby and Peter Dolman — see Bibliography. Mills in isolated places frequently have the name of the nearest settlement attached to them although they may not stand in that parish. The type of mill is indicated: post mill (p); smock mill (s); tower mill (t); some visible remains (x).

AVON
Brockley (t); Clifton Down, Bristol (t); Falfield (t); Felton Common (t); Frampton Cotterell (t); Hutton (t); Kenn (t); Locking (t); Portishead (t) — now part of golf club house; Uphill (t); Warmley (t); Worle Hill (t); Worle Vale (t).

BEDFORDSHIRE
Dunstable (t); Sharnbrook (t); Stanbridge (t); Stevington (p); Totternhoe (t) — Dolittle Mill, a combined windmill and watermill.

BUCKINGHAMSHIRE
Bradwell, New (t); Brill (p); Cholesbury (t); Coleshill (t); Ibstone (s); Lacey Green (s); Pitstone (p); Quainton (t); Wendover (t).

CAMBRIDGESHIRE
Barnack (t); Bourn (p); Buckden (t); Great Chishill (p); Eaton Socon (t); Histon (s); Ickleton (t); Kneesworth (t); Little Wilbraham (t); Madingley (p); Over (t); Six Mile Bottom (p); Soham (t); Steeple Morden (s); Stretham (t); Swaffham Prior (t); Swavesey (t); West Wratting (s); Wicken Fen (s).

CHESHIRE
Great Saughall (t); Neston (t); Upton-by-Chester (t); Willaston (t).

CLEVELAND
Elwick (t); Hart (t).

CORNWALL
Carlyon (t); Mount Hermon (t).

CUMBRIA
Cardewlees (t); Cockermouth (t); Haverigg (t); Langrigg (t); Monkhill (t); Wigton (t); Workington (t).

DERBYSHIRE
Dale Abbey (Cat & Fiddle Mill) (p); Heage (t); Kirk Hallam (p); Normanton (t); Bolsover (t).

DEVON
Broadclyst (Cliston Manor) (t); Galmpton Warborough (t); Instow (t); North Whilborough (t); Paignton (Fernacombe) (t); Petrockstow (Heanton) (t); Torquay (Yaddon Down Mill, Audley Avenue) (t). All remains without sails.

GAZETTEER

DORSET
Shaftesbury (t) — built 1969 in the Portuguese style.
DURHAM
Aycliffe (t); Easington (t); Ferryhill (t); Hutton Henry (t).
ESSEX
Ashdon (p); Aythorpe Roding (p); Bocking (p); Clavering 2
mills (t); Debden (t); Dunmow (t); Finchingfield (p); Gainsford
End (Toppesfield) (t); Great Bardfield (t); Mountnessing (p);
Ramsey (p); Rayleigh (t); Stansted Mountfitchet (t); Stock (t);
Terling (s); Thaxted (t); Tiptree (t); White Roding (t).
GUERNSEY, C.I.
St Martin's (t).
GWYNEDD
Felin Adda (t); Felin Llanerchymedd (t); Llyn On, Llan-
deusant (t); Melin-y-Bont, Bryn Du (t) — a windmill with a
waterwheel and a list of miller's charges in Welsh.
HAMPSHIRE
Bursledon (t); Chalton (t); Langstone (t); Portchester (t).
HEREFORD AND WORCESTER
Avoncroft Museum of Buildings, Bromsgrove (p).
HERTFORDSHIRE
Cromer (p); Croxley Green (t); Tring (t).
HUMBERSIDE
Bainton (t); Barton-on-Humber 4 mills (t); Beverley 2 mills
(t); Bridlington (t); Burstwick (t); Ellerton (t); Goole (t); Hessle
(t); Hibaldstow (t); Howden (t); Keyingham (t); Kirton Lindsey
(t); Nafferton (t); Ousefleet (t); Patrington (t); Scawby (t);
Scunthorpe (t); Seaton Ross 2 mills (t); Skidby (t); Stallingbor-
ough (t); Swinefleet 2 mills (t); Waltham (t) — once 6 sails;
Wrawby (p); Wressle (t); Yapham (t); Yokefleet (t).
ISLE OF WIGHT
Bembridge (c 1700) (t).
KENT
Benenden (s); Bidborough (t); Canterbury (1817) — St
Martin's Hill (t); Charing (s); Chillenden (1868) (p); Chislet (s);
Cranbrook — Union Mill (1814) (s); Eastry (s); Edenbridge (t);
Guston (1849) — Swingate Mill (t); Herne (1781) (s); Margate
— Draper's Mill (s); Meopham (1801) (s); Northbourne (1848)
(s); Oare (t); Ripple (s); Rolvenden (p); Sandwich — White Mill
(s); Sarre (1820) (s); Stanford (t); Stelling Minnis (s); St
Margaret's Bay (1929) (s); West Kingsdown (s); Whitstable
(1815) — Boarstall Hill (s). Willesborough (1869) (s); Wittersham (1781) — Stocks Mill (p); Woodchurch — Lower Mill (s).
LANCASHIRE (including Greater Manchester)
Bickerstaff (1756) (t); Bretherton (1741) (t); Clifton (1790)
(t); Haigh (1875) (t); Holmeswood (1850) (t); Kirkham (c 1780)
(t); Little Marton (1838) (t); Lydiate (c 1768) (t); Lytham (1805)
(t); Parbold (1817) (t); Pilling (1808) (t); Preesall (1839) (t);

Preston (t); Staining (t); Thornton (1794) (t); Treales (t); Wrea Green (t); Wrightington (t).

LEICESTERSHIRE

Kibworth Harcourt (p); Morcott (t); Waltham-on-the-Wolds (t); Whissendine (t); Wymondham (t) — once 6 sails.

LINCOLNSHIRE

Alford (t) — 5 sails; Baston (t); Boston (Maud Foster mill) (t) — 5 sails; Bourne (x); Burgh-le-Marsh (t) — 5 sails; Heapham (t); Heckington (t) — 8 sails; Holbeach (t); Huttoft (t); Kirkby Green (t); Kirton End (t); Langtoft (t); Lincoln (t); Long Sutton (t) — 6 sails; Mablethorpe (x); Sibsey (t) — 6 sails; Sleaford (t); Stickford (t); Stickney (t); Swineshead (t); Wainfleet (t); Wigtoft (t); Winthorpe (t); Wragby (t).

LONDON

Arkley (t); Brixton (t); Plumstead Common (t); Shirley (s); Upminster (s); Wandsworth Common (s); Wimbledon Common (hollow-post mill — originally).

MERSEYSIDE

Bidston (t); Gayton (t); Great Crosby (1813) (t).

NORFOLK

The drainage mills are indicated by the additional letter (d).
Berney Arms (dt); Billingford (t); Burnham Overy (t); Caston (t); Cley (t); Denver (t); Garboldisham (p); Great Bircham (t); Halvergate (dt); Hickling (t); Hindringham (t); Horning (dt); Horsey (dt); How Hill (2 dt); Hunsett (dt); Little Cressingham; Old Buckenham (t); Paston (t); Potter Heigham (dt); Ringstead (t); Runham Swim (t); St Benet's Abbey (dt); St Benet's Level (dt); St Olaves (s); Stalham (dt); Starston (d hollow-post); Stracey Arms (dt); Sutton (t); Thurne Dyke (dt); West Winch (t); Weybourne (t); Wicklewood (t); Worstead (t).

NORTHAMPTONSHIRE

Greens Norton (t); Hellidon (t).

NORTHUMBERLAND

Acomb (t); Bamburgh (t); Chollerton (t); Great Whittington (t); Haggerston (t); Hartley (t); High Callerton (t); Scremerston (t); Spindlestone (t); Woodhorn (t).

NOTTINGHAMSHIRE

North Leverton (t); Nottingham (Sneinton) (t).

OXFORDSHIRE

Blackthorn (t, x); Bloxham (p); Milton Common (t); North Leigh (t); Wheatley (t).

SOMERSET

Chapel Allerton (t); Curry Rivel (t); High Ham (t); Shapwick (t); Stone Allerton (t); Walton (t); Watchfield (t); West Monkton (t).

SUFFOLK

Dalham (s); Drinkstone (p); Framsden (p); Friston (p); Great Thurlow (s); Herringfleet (s); Holton (p); Pakenham (t);

GAZETTEER

Saxtead Green (p); Stowmarket, Museum of East Anglian Life (s); Syleham (p); Thelnetham (t); Thorpeness (p); Woodbridge — (2 t).

SURREY

Ewhurst (t); Frimley Green (t); Outwood (p); Reigate (Wray Common) (t); Reigate Heath (p); Tadworth (p).

SUSSEX, EAST

Alfriston (t); Chailey (North Common) (s); Cross in Hand (p); Herstmonceux (p); Icklesham (p); Mayfield (Argos Hill) (p); Mark Cross (t); Nutley (p); Patcham (t); Polegate (t); Punnett's Town (s); Rottingdean (s); Rye (s); Stone Cross (t); West Blatchington (s); Winchelsea (p).

SUSSEX, WEST

Angmering (t); Arundel (t); Barnham (t); Clayton 'Jack' (t), 'Jill' (p); Earnley (s); East Wittering (t); Gatwick Manor (s); Halnaker (t); High Salvington (p); Keymer (p); Lowfield Heath (p); Nutbourne (t); Pagham (Nyetimber) (t); Selsey (t); Shipley (s); Washington (s); West Chiltington (s); Weald and Downland Open Air Museum, Singleton — hollow-post mill.

TYNE AND WEAR

Cleadon (t); Fulwell (t); Newcastle (Chimney Mill) (t); Heaton Park (t); North Shields (t); West Boldon (t); Whickham (t); Whitburn (t).

WARWICKSHIRE

Chesterton (t); Kenilworth (t); Napton (t); Norton Lindsey (t); Packwood (t); Thurlaston (t); Tysoe (t).

WEST MIDLANDS

Balsall Common (t).

WILTSHIRE

Wilton (near Great Bedwyn) (t).

YORKSHIRE, NORTH

Appleton Roebuck (t); Askham Richard (t); Elvington Brickyard (t) — a pumping mill; Kellington (t); Kirkbymoorside (t); Riccall (t); Skelton (t); South Duffield (t); Stutton (t); Tollerton (t); Ulleskelf (t); York (Holgate) (t).

YORKSHIRE, SOUTH

Branton (t); Fishlake (t); Hatfield (t); Hatfield Woodhouse (t); Moss (Wrancarr) (t); Sykehouse (t); Thorne (t); Wentworth (2 t).

YORKSHIRE, WEST

Aberford (2 t); Birstall (t); Bramham (t); Darrington (t); Kippax (t); Pontefract (t); Swillington (Colton) (t).

NORTHERN IRELAND - COUNTY DOWN

Ballycopeland (t).

13. Bibliography

Brangwyn, F. and Preston, H. *Windmills* 1923

Batten, M. I. *English Windmills* Vol. 1. Architectural Press 1930

Dolman, Peter C. J. *Windmills in Suffolk* Suffolk Mills Group 1978

Farries, K. G. and Mason, M. T. *Windmills of Surrey and Inner London* Skilton 1966

Flint, Brian *Suffolk Windmills* Boydell Press 1979

Freese, Stanley *Windmills and Millwrighting* David and Charles 1971

Hemming, Peter *Windmills in Sussex* 1936

Hopkins, R. Thurston and Freese, Stanley *In Search of English Windmills* 1931

Hughes, J. *Cumberland Windmills* Cumberland and Westmorland Archaeological Society 1972

McDermott, Richard and Richard *The Standing Mills of West Sussex* 1978, *The Standing Mills of East Sussex* 1978, Betford

Major, Kenneth *The Mills of the Isle of Wight* Skilton 1970

Minchington, Walter *Windmills of Devon* Univ. of Exeter (Dept. of Econ. Hist.) 1977

Scott, Martin *The Restoration of Windmills and Windpumps in Norfolk* Norfolk Windmills Trust 1977

Seaby, Wilfred A. and Smith, Arthur *Windmills in Warwickshire* Warwickshire Museum 1977

Smith, Arthur *Windmills in Hertfordshire* 1974, *Windmills in Bedfordshire* 1975, *Windmills in Cambridgeshire* 1975, *Windmills in Buckinghamshire and Oxfordshire* 1976, *Windmills of Surrey and Greater London* 1976, *Windmills in Huntingdonshire and Peterborough* 1977, *Drainage Windmills of the Norfolk Marshes* 1978, *Windmills in Sussex* 1980 Stevenage Museum Publications.

Smith, Donald *English Windmills* Vol. II Architectural Press 1932

Turpin, B. J. and J. M. *Windmills in Kent* 1979, *Windmills in Essex* 1977, Windmill Publications

Vince, John *Windmills* Basil Blackwell 1975

Vince, John *Windmills in Buckinghamshire and the Chilterns* 1976. Available from the author

Vince, John *Discovering Watermills* Shire 1980

Vince, John, *Old Farms* John Murray 1982

Vince, John *Power Before Steam* John Murray 1985

Wailes, Rex *The English Windmill* 1954

Wailes, Rex *Windmills in England* Skilton 1976

Watts, Martin *Somerset Windmills* Agraphicus 1976

West, Jenny *The Windmills of Kent* Skilton 1971

Windmills to Visit Norfolk Windmills Trust.

Index of mills